Population and Biodiversity

POPULATION AND BIODIVERSITY

By

Dr. M. Lakshmi Narasaiah
M.A., Ph.D.
Professor of Economics,
Co-ordinator, Department of M.B.A. and Commerce,
Special Officer,
Sri Krishnadevaraya University Post-graduate Centre,
Kurnool–518 002
Andhra Pradesh (India)

DISCOVERY PUBLISHING HOUSE
NEW DELHI

Published by:
Namit Wasan

DISCOVERY PUBLISHING HOUSE PVT. LTD.
4383/4B, Ansari Road, Darya Ganj
New Delhi-110 002 (India)
Phone : +91-11-23279245; 23253475; 43596065
E-mail : discoverybooksindia@gmail.com
discoverypublishinghouse@gmail.com
namitwasan9@gmail.com
web : www.discoverypublishinggroup.com

First Published: **2006**

Reprinted: **2021**

ISBN: 978-81-8356-065-8

Population and Biodiversity

Printed at:
Infinity Imaging Systems
Delhi

Preface

As human population has surged this century, the populations of numerous other species have tumbled, many to the point of extinction. Indeed, we live amid the greatest extinction of plant and animal life since the dinosaurs disappeared some 65 million years ago, with species losses at 100 to 1,000 times the natural rate. But humans are not just witnesses to a rare historic event, we are actually its cause. The leading sources of today's species loss, habitat alteration, invasions by exotic species, pollution, and overhunting are all a function of human activities.

Human activities have pushed the percentage of mammals, amphibians, land fish that are in "immediate danger" of extinction into double digits. The principal cause of species extinction is habitat loss—the result of encroachment by humans for settlements, for agriculture, or to claim resources such as timber. A particularly productive but vulnerable habitat is found in coastal areas, home to 60 per cent of the world's population. Coastal wetlands nurture two-thirds of all commercially caught fish, for example. And coral reefs have the second highest concentration of biodiversity in the world, after tropical rainforests. But human encroachment and pollution are degrading these areas: roughly half of the world's salt marshes and mangrove swamps have been eliminated or radically altered, and two-thirds of the world's coral reefs have been degraded, 10 per cent of them "beyond recognition". As coastal migration continues—coastal dwellers could account for 75 per cent of world population within 30 years—the pressures on these productive habitats will likely increase".

Habitat loss tends to accelerate with an increase in a country's population density. This is bad news for the world's biodiversity hotspots-species-rich ecosystems at greatest risk of destruction. Twenty-four of these hotspots, containing half of the planet's species, have been identified globally. Some of the most important hotspot countries will reach population densities that have been linked with very high rates of habitat loss. Five of the six most biologically rich countries could see more than two-thirds of their original habitat destroyed by 2050 if this historical relationship holds.

Related to loss of habitat is the growing incidence of plant, animal, insect, and microbial invasions of ecosystems worldwide as human interchange increases. These "exotic species" sometimes dominate local ecosystems, eliminating native species and reducing overall diversity. Exotics are implicated in 68 per cent of all fish extinctions in the United States this century, for example. Growth in human travel and commerce explains many accidental invasions by exotics, but foreign species are also deliberately introduced into farms, plantation forests, and aquaculture systems. Although only 1 per cent of exotics cause widespread damage, exotic species are the second leading cause, after habitat destruction, of species loss worldwide.

Dr. M. Lakshmi Narasaiah

Contents

	Preface	*v*
1.	Fertility Rates: The Decline is Stalling	1
2.	The Good News about Population Growth	5
3.	Population Growth Facts and Figures	10
4.	Population and the Environment: The Global Challenge	13
5.	Measuring Population's Impact	16
6.	The Population Challenge	19
7.	What is Known about Reducing Maternal Mortality?	25
8.	Safe Motherhood is a Human Rights Issue	27
9.	Ecosystems, Our Unknown Protectors	30
10.	Forests: The Earth's Lungs	33
11.	Biodiversity	37
12.	Living with Diversity	40
13.	Global Warming: Worrisome Signs	44
14.	Climate Change	49
15.	Forests	52
16.	An Agenda for Change	55
17.	Ecotourism or Ecocide?	59
18.	Urbanisation and the Environment	64
19.	Towards Healthy Cities	68

20. Sustainable Cities 72
21. Consuming the Future 76
22. The Future of Work 79
23. Energy and Sustainability 85
24. Development: The Third Way 89
25. Employment and Promoting Ecology: How a Service Culture Could Put People Back to Work 93
26. South Asia Quarrels Over Water 99
27. Using Economics to Advantage 102
28. A Crucial Encounter 105
29. Sustainable Tourism and the Environment 108
30. Pro-Poor Tourism: Opportunities for Sustainable Local Development 113
31. Consumption Bomb 122
32. The Persistence of Indian Poverty and its Alleviation 126
33. Employment and Poverty Alleviation 134
34. Food Production 140
35. No Progress without a Secular Society 144
36. What's Driving Migration 147
37. Major Cyclones in Andhra Pradesh: Some Observations 153
Bibliography 159
Index 163

1

Fertility Rates

The Decline is Stalling

During the 1970's, one of the population trends was the reduction in the total fertility rate in several key countries, including the world's two largest nations—China and India. (the fertility rate measures the average number of children born to women in their childbearing years.) In China, the rate dropped precipitously, from 6.4 children per woman in 1968 to 2.2 in 1980. In India, the decline was more modest, but still significant: from 5.8 children per woman between 1966 and 1971 to 4.8 children 1976 and 1981.

These trends helped slow the rate of world population growth from 2.1 per cent between 1965 and 1970 to 1.7 per cent between 1975 and 1980. At that point, however, the decline in the number of children that women were having in these two population giants stalled.

In China, despite the most aggressive and least democratic population control program in the world, the fertility rate remained around 2.5 throughout much of the 1980s as couples continued to want to marry young and to have two or more children. In India, the overzealous promotion of family planning by the ruling Congress Party through 1977 apparently backfired after the party's defeat and progress toward lower birth rates ran out of steam.

One important lesson from these experiences is that governments must do more than just supply contraceptives;

they need to lower the demand for children by making fundamental changes that improve women's lives and increase their access to and control over money, credit, and other resources.

Many countries still register fertility rates above replacement level (See Table), which is generally 2.1 children per woman or basically two children per couple. The total fertility rate for the world as a whole in 1995 was 3.3 ranging from 1.8 in more developed nations to 4.4 in less developed one (excluding China). In a number of countries, such as Brazil, Egypt, Indonesia, Mexico, and Thailand, fertility rate have been dropping as they did in the 1970s in China and India. At the same time, developing countries have not yet entered the demographic transition.

The demographic transition occurs when both birth rates and death rates in a country drop from historically high levels to low ones that translate into a stable population—one that merely replaces itself with each new generation. Traditionally, although not always, death rates have declined first, following the spread of sanitation and improved health care coverall. Rapid population growth often follows this first phase of the demographic transition, as the gap between fertility and mortality for a time. Eventually, however, fertility rates fall too.

As the moment, they remain high a number of countries. The reasons include unequal rights and opportunities for women, as well as inadequate access to birth control. Whatever the reason, the effect is the same: 67 countries, home to 17 per cent of the world population, are at best in the early stages of a transition to low fertility rates. Most of them are in Africa and South Asia, and their population is likely to double in 20 to 25 years.

This is leading to a two-tiered demographic world that is every bit as worrying as the world of economic haves and have-nots. Countries such as Nigeria and Pakistan are finding it harder to keep up with the demand for food, health

Population Size, Fertility Rate, and Doubling Time, 20 Largest Countries, 1995

Country	*Fertility Population (millions)*	*Rate (average number of children per woman)*	*Doubling Time (years)*
Italy	58	1.3	3466
Germany	81	1.4	*
Japan	125	1.5	217
United Kingdom	58	1.8	267
France	58	1.8	169
Russia	149	1.7	990
United States	258	2.0	92
China	1178	1.9	60
Thailand	57	2.4	49
Indonesia	188	3.0	42
Brazil	152	3.6	46
Turkey	61	3.6	32
Mexico	90	3.4	13
India	97	3.9	34
Vietnam	72	4.0	31
Philippines	5	4.1	28
Egypt	8	4.6	30
Pakistan	122	6.7	23
Iran	3	6.6	20
Nigeria	5	6.6	23

Source: Population Reference Bureau, 1995 World Population Data Sheet (Washington, DC, 1995).

care, jobs, housing, and education than countries that are in the middle of the demographic transition.

Even when a country does reach replacement-level fertility, its population can continue growing for decades. There is a built-in momentum created by all the people who have yet to enter their childbearing years. Indeed, the decline in the world's population growth rate stalled in the 1980s in part because even in China, India, and other countries where fertility rates had been dropping, large number of people who had been born in the 1960s reached childbearing age. So even if couples had two or three children instead of five or six, their parents did, the population would grow substantially.

For the world as a whole, even if replacement—level fertility had been achieved in 1990, the population would continue to grow until it reached 8.4 billion in 2150 because of all the young people already alive. This built-in momentum obviously limits how quickly and country can stop population growth. Nevertheless, reaching replacement level fertility is an all-important first step. The 67 countries that have not yet begun the demographic transition-nations in which invariably the government believes fertility levels are to high—could move in the right direction by providing the contraceptive and health care services that would help couples have only the number of children they desire.

2

The Good News about Population Growth

Global population is growing at an ominous pace. Yet new opportunities, if they are seized, along with changes in government policies, could help solve the population problem within two generations. Three facts of the impact of population growth on economic and human development are:

- The countries of the developing world already contain three-quarters of humanity and will soon be home to nine of every 10 people on Earth.
- With but a small fraction of the world's GNP, a tiny percentage of its scientists, the lion's share of its debt and nearly all of its abject poor, these countries have been ill-equipped to absorb the dramatic increases in human numbers that, despite the steadily declining rate of global population growth, has occurred since the end of World War III.
- Within the next 60 years, between three and seven billion more people will be added to the population of the developing world. Assuming they can sustain such increases, developing countries will double or triple in size before their population are stabilized a century or more from now. If they cannot, and population size outstrips the resources available to support it, population growth will be slowed faster and sooner, on nature's grim terms.

The conventional wisdom is that because rapid population growth lowers per capita expenditures on education, per capita access to renewable natural resources, and per capita savings rates, it retards economic and social development. But the problem goes beyond this conventional wisdom. There is a lethal synergy that exists in dozens of developing countries between the sheer speed of population growth, on the one hand, and inappropriate government policies that magnify its negative effects, on the other. The former is suggested by the fact that the populations of most developing countries are projected to double in size in 30 years or less. The later is evident in policies that have prevented many developing countries from either slowing such growth or accommodating it gracefully.

It would stretch the capacities of even the most prosperous countries if they had to accommodate the kind of population growth expected in regions including Sub-Saharan Africa and Sough Asia. Without the benefit of prosperity, the indispensable requirement will be for governments to formulate policies that will alleviate rather than exacerbate the economic and social effects of rapid population growth. But as examples from around the developing world illustrate, such policies are in short supply. In Africa, food supplies are inadequate partly because of a pervasive urban basis that leaves farmers with few incentives to produce. In dozens of command economies, housing for low-income families is limited because of anachronistic rent control laws, which have discouraged private developers from building new residential housing units. In other developing countries, significant resources that could be invested in education are diverted for military use. In few of these countries is rapid population growth the main factor retarding economic growth. But in most, such growth has exacerbated the negative effects of short-sighted policies and it will likely do so to a greater degree in the future, for reasons ranging from inexperience in self-government to political opportunism to bureaucratic ineptitude.

High political and institutional barriers thus separate what can be done to mitigate the effects of rapid population growth from what is likely to be done in all to many countries. That's the bad news one has to face as they grapple with the problem of rapid populatin growth. The good news is that where such barriers have been lowered by far-sighted governments, the results have been hopeful and in some cases dramatic. Sustained high agricultural growth has taken place, in a number of countries—where constraints on small-scale agriculture have been removed. Significant progress has also been achieved by some of the thousands of non-governmental organisation (NGOs) that are playing an increasingly important role in solving population-related problems at the local level in developing countries around the world.

In Bangladesh, meanwhile, the innovative work of two other well-known NGOs – the Grameen Band and the Bangladesh Rural Advancement Committee – has helped to unleash the productive potential of scores of thousands of poor women, with clear implications not only for economic development but also for fertility reduction. Researchers say the "micro-enterprise" loans given by the banks have changed attitudes towards childbearing by empowering women, raising the "opportunity costs" of having large families and producing a greater degree of parity in marriage relationships. Creating a sense of a future that extends beyond mere day-to-day survival has changed the context for decisions about family size.

Nor is the work of NGOs the only hopeful development weighing in the balance against unprecedented increases in population that are all but certain to occur over the next few decades. As the 21st century approaches there is a heartening convergence of factors that augur well for stabilizing global population, sooner perhaps than experts would have thought possible even a decade ago.

The most consequential factor many be the growing desire on the part of men and women around the world for access to family planning services. Millions more, including sexually active young people and couples discouraged from practising family planning by the poor quality of existing services, could be reached by well-run family planning programmes, especially those sensitive to the reproductive health needs of women. Family planning agencies are confident that, in the process of satisfying the existing demand for family planning services, they can create new demand by improving the quality of services and legitimizing the small-family model.

Across the institutional spectrum, from national governments in rich poor nations alike, to giant multilateral institutions there is a growing recognition that the future will pose far greater challenges to the human race if rates of population growth are not slowed further. Significantly, that conclusion has recently been endorsed by a growing number of natural scientists who now warn that, without global efforts to slow population growth, science and technology may not be able to redeem the future from want and hunger. One can predict that if current population and consumption trends continue, "science and technology may not be able to prevent either irreversible degradation of the environment or continued poverty for such of the world".

Rates of contraceptive use among men and women, who in every other respect regard themselves as faithful Catholics, are as high as or higher than those of non-Catholics. The lowest birthrates and highest contraceptive use rates on record have been achieved in Catholic World, where abortion is also widely favoured as a back-up in case of contraceptive failure. Similar attitudes are found in the Islamic World, where rapid population growth has prompted theologians to accent a more permissive side of their theology with respect to the use of modern contraceptives. The results have been strong clerical support for family planning programmes.

Beyond the diminished obstacles to progress is the increased opportunity for progress that has been created by the end of the cold war. Freed from the necessity of concentrating on geostrategic threats to peace, policy makers now have the luxury of turning their attention on the global forces that impinge on the peace and prosperity of nations, including the pressure population growth is placing on economic development, food supplies and the natural environment. Worried that such pressures could destabilize the governments of poor nations, prompt interstate conflict over scarce resources, and spur disruptive movements of populations into urban areas or across national frontiers, Western analysts are even now thinking of ways to retool foreign policy to cope with the proliferating number of local threats. One crucial task will be to monitor environmental and economic conditions around the developing world and to provide early warning of regional conflicts. More difficult to implement will be longer-range approaches, including a refashioning of the international trading system to provide more equity to poor nations.

If political leaders are willing and able to capitalize on the opportunity provided by the coming together of this extra-ordinary constellation of circumstance, the population problem could be solved within two generations. If they are not, providing for the welfare of humanity in the century ahead will be more difficult; perhaps, in some of the world's poorest regions, it will prove impossible.

3

Population Growth Facts and Figures

The world-wide rate of population growth has been essentially the same since 1975, and about 1.7 per cent a year. Fertility is actually going down slightly, from 3.8 in 1975-1980 to 3.3 in 1990-1995. Because of past growth, however, the number of people added each year is still rising. In 1975, the annual addition was about 72 million. In 1992 it was 93 million. It will peak between 1995 and 2000, at about 98 million annually.

Rapid population growth is therefore still the dominant feature of global demographics, and will continue to be sc for atleast the next 30 years. The 1993 global population of 5.57 billion is projected to increase to 6.25 billion in 2000, 8.5 billion in 2025 and 10 billion in 2050; significant growth will probably continue until about 2150 and a level of about 11.6 billion.

The developing countries proportion of this increase grew from 77 per cent in 1950 to 93 per cent in 1990; between now and the end of the century it will be 95 per cent. Africa and South Asia alone account for 53 per cent.

Asia's population in mid-1993 was 3.3 billion. By 2025, Asia will have 4.9 billion people, equal to the whole of world population in 1986; Africa (700 million) will have 1.6 billion; Latin America (466 million) will have 700 million people.

This overall picture conceals wide variations from country to country and region to region. For example:

- Annual growth 1990-1995 is estimated at 3.0 per cent for Africa, with Asia at 1.9 per cent and Latin America 2.1 per cent. By the large the fastest rates of growth are in the poorest countries.
- The 47 countries officially designated by the demographers as "least developed" accounted for 7 per cent of global increase in 1950, but 13 per cent by 1990.
- Life expectancy has increased by 30 years in East Asia over the last four decades, as against 15 years in Africa, which has 30 of the 47 "least developed" countries.
- Fertility has fallen by 60 per cent in East Asia in the same period, but by only 25 per cent in South Asia and hardly at all in Africa.
- Maternal mortality has been halved in East Asia, but remains virtually unchanged in South Asia and Africa.

Among developing countries, the lowest growth rates are in East Asia and the Caribbean (1.3 per cent). East Asia's growth rates largely reflect the situation in China, which is 85 per cent of the region. Central and South America, South-East and South Asia and Southern Africa lie between 2 and 2.5 per cent; North Africa and West Asia between 2.5 and 3; and the rest of Africa over 3 per cent.

The biggest variation of all has grown up between the industrialized countries of Europe and North America, and the rest of the world. In the industrialized countries population growth has slowed or stopped altogether, and fertility is at or below replacement level. Their populations increased by 43 per cent between 1950 and 1990, compared with 162 per cent among the least developed countries and 140 per cent in the other developing countries. This variation will deepen: the populations of Europe and Sub-Saharan Africa, roughly the same in 1985 at about 480 million, will be 500 million and 1500 million respectively by 2025.

Asia has 59 per cent of world population, Latin America 9 per cent and Africa 12 per cent. Africa's share is projected to go up to 19 per cent by 2025, while proportions in the other regions remain about the same. Within Asia the proportions are chaning: China, now 37 per cent of Asia's population, will be 31 per cent by 2025; India will go from 27 per cent to 29 per cent.

4

Population and the Environment

The Global Challenge

As the century begin, natural resources are under increasing pressure, threatening public health and development. Water shortages, soil exhaustion, loss of forests, air and water pollution, and degradation of coastlines afflict many areas. As the world's population grows, improving living standards without destroying the environment is a global challenge.

Most developed economies currently consume resources much faster than they can regenerate. Most developing countries with rapid population growth face the urgent need to improve living standards. As we humans exploit nature to meet present needs, are we destroying resources needed for the future?

Environment Getting Worse

In the past decade in every environmental sector, conditions have either failed to improve, or they are worsening:

Public Health. Unclean water, along with poor sanitation, kills over 12 million people each year, most in developing countries. Air pollution kills nearly 3 million more. Heavy metals and other contaminants also cause widespread health problems.

Food Supply. Will there be enough food to go around? In 64 of 105 developing countries, studied by UN Food and Agricultural Organisation, the population has been growing faster than food supplies. Population pressures have degraded some 2 billion hectares of arable land—an area the size of Canada and the US.

Fresh Water. The supply of fresh water is finite, but demand is soaring as population grows and use per capita rises. By 2025, when world population is projected to be 8 billion, 48 countries, containing 3 billion people will face shortages.

Coastlines and Oceans. Half of all coastal ecosystems are pressured by high population densities and urban development. A tide of pollution is rising in the world's seas. Ocean fisheries are being over-exploited, and fish catches are down.

Forests. Nearly half of the world's original forest cover has been lost, and each year another 16 million hectares are cut, bulldozed, or burned. Forests provide over US$400 billion to the world economy annually and are vital to maintaining healthy ecosystems. Yet, current demand for forest products may exceed the limit of sustainable consumption by 25 per cent.

Biodiversity. The earth's biological diversity is crucial to the continued vitality of agriculture and medicine—and perhaps even to life on earths itself. Yet human activities are pushing many thousands of plant and animal species into extinction. Two of every three species is estimated to be in decline.

Global Climate Change. The earth's surface is warming due to greenhouse gas emissions, largely from burning fossil fuels. If the global temperature rises as projected, sea levels would rise by several meters, causing widespread flooding. Global warming also could cause droughts and disrupt agriculture.

Toward a Livable Future

How people preserve or abuse the environment could largely determine whether living standards improve or deteriorate. Growing human numbers, urban expansion, and resource exploitation do not bode well for the future. Without practicing sustainable development, humanity faces a deteriorating environment and may even invite ecological disaster.

Taking Action. Many steps toward sustainability can be taken today. The include using energy more efficiently; managing cities better; phasing out subsidies that encourage waste; managing water resources and protecting fresh-water sources; harvesting forest products rather than destroying forests; preserving arable land and increasing food production through a second Green Revolution; managing coastal zones and ocean fisheries; protecting biodiversity hotspots; and adopting an international convention on climate change.

Stabilizing Population. While population growth has slowed, the absolute number of people continues to increase 0 by about 1 billion every 13 years, slowing population growth would help improve living standards and would buy time to protect natural resources. In the long run, to sustain higher living standards world population size must stabilize.

5

Measuring Population's Impact

There is no easy way to measure the overall impact of human activities on the environment. Nevertheless, several approaches have been developed as follows:

Environmental Resource Accounting

Environmental resources accounting attempts to place an economic value on "environmental goods and services" used—natural resources that conventionally have been regarded as free and used in common. These include unpolluted freshwater, clean air, ocean life, forests, and wetlands.

Some economists argue that the value of environmental goods and services should be incorporated into estimates of Gross manufactured capital, which depreciates in value overtime, environmental capital (such as forests fisheries, and unpolluted air and water) currently is not considered to depreciate, and no charge is made against current income as it is used. A country could exhaust its mineral resources, cut down its forests, erode its soils, pollute its acquires, and hunt its wildlife and fisheries to extinction, but measured income would not be affected as these natural assets disappeared.

If natural resources were valued in the same way that manufactured assets are valued, it might help economies learn to use them more efficiently and to conserve them in order to ensure continued use in the future. Such valuations also might help indicate the economic benefits of protecting the environment, as well as the ecological benefits. In other terms, instead of continuing to draw down their "environmental capital" until it is gone, economies could being to live on its interest, maintaining the capital for use indefinitely in the future.

I = P × A × T

The equation I = P × A × T represents another effort to describe the overall impact of humanity on the environment. In the equation:

I is environmental impact

P is population (including size, growth, and distribution)

A is the level of affluence (consumption per capita), and

T is the level of technology.

Despite its limitations—for instance, inability to assign actual values to each component or to depict changes in the factors over time—the equation is valuable. In particular, it emphasizes that developing countries with large and rapidly growing populations affect the environment, even though their levels of affluence may be low, while at the same time countries in the developed world with little or no population growth have a substantial environmental impact because consumption per capital is so high.

The equation makes clear that slowing population growth is a key part of any strategy to reduce humanity's impact on the environment. For example, even if per capita resource consumption (A) declined or technologies (T) improved enough to reduce the environmental impact (I) of humanity by 10 per cent, this gain would be wiped out in less than a decade because world population (P) is growing at over 1 per cent per year since per capita consumption of resources is expected to increase as living standards rise, protecting the environment requires more efficient production technologies, less waste, and ultimately a stable world population size.

Ecological Footprints of Nations

Every body has an impact on the Earth, because they consume the products and services of nature. Their ecological impact corresponds to the amount of nature they occupy to keep them going. In other words, we calculate the 'ecological footprints' of these countries.

Carry Capacity

The term "carrying capacity" refers to the number of people the earth can support. Logically, population growth must stop at some point, to the earth would become overcrowded and its resources eventually would be depleted. But what is this maximum human population?

This question has been debated since 1798. When English economists Thomas Malthus predicated that population growth inevitably would outstrip the food and water supply at some point. Since the estimates of carrying capacity have varied a great deal depending on what assumptions are made out technology, consumption levels, and other factors that are not easily forecast. Some have even argued that the earth's carrying capacity may already have been exceeded in sense that the world consumed at the rate that Americans and Western European consume.

While no body can know how many people the earth could support, few would want to find out the hard way—by reaching this theoretical limit. Calculate the maximum number of people who could exist on earth seems less important than determining how resources can be used wisely and managed sustainably to improve living standard without eventually destroying the natural environment that

6

The Population Challenge

During the last half-century world population has more than doubled, climbing from 2.5 billion in 1950 to 5.9 billion in 1998. Those of us born before 1950 are members of the first generation to witness a doubling of world population. Stated otherwise, there has been more growth in population since 1950 than during the 4 million years since our early ancestors first stood upright.

This unprecedented surge in population combined with rising individual consumption, is pushing our claims on the planet beyond its natural limits. Water tables area falling on every continent as demand exceeds the sustainable yield of aquifers. Eventual aquifer depletion will bring irrigation cutbacks and shrinking harvests. Our growing appetite for seafood has pushed oceanic fisheries to their limits and beyond. Collapsing fisheries tell us we can go no further. The Earth's temperature is rising, promising changes in climate that we cannot even anticipate. We are triggering the greatest extinction of plant and animal species since the dinosaurs disappeared. As our numbers go up, their numbers go down.

These effects of population growth are relatively recent, but assertions that population growth could affect human welfare are not. In 1798 Thomas Malthus, a British clergyman and intellectual, warned in his famous piece, *An Essay on the Principles of Population,* of the tendency for population to grow exponentially while food supply grew arithmetically. He saw a world where human numbers would continually press against available food supplies.

During the 200 years since Malthus issued his warning, famine has visited countries as diverse as Ireland and India, Ethiopia and China. Indeed, despite the near-tripling of the world grain harvest since 1950 the hungry and malnourished in 1998 number an estimated 840 million—nearly as many people as lived in the world when Malthus penned his essay.

But the nature of famine has changed. Whereas it was once geographically defined by areas of poor harvests, today famine is economically defined by low incomes in those segments of society that lack the purchasing power to buy enough food. Famine concentrated among the poor is less visible than the more traditional version but is no less real.

In addition to checks imposed by food shortages, there is evidence that other checks on population growth are now emerging, such as new infectious diseases, including AIDS, Ethnic conflicts within societies, such as Rwanada and the Sudan, are also taking a growing toll. Water shortages on a scale that would deprive people of enough water to produce food could undermine governments.

The evidence gathered here indicates that the rapid population growth prevailing in a majority of the world's countries is not going to continue much longer. Either countries will get their act together, shifting quickly to smaller families, or death rates will rise from one or more of the stresses just mentioned. As human demands press against more and more of the Earth's limits, the questions is not whether population growth will slow, but how. Will it be because countries do it humanely by shifting quickly to smaller families? Or because they fail to do so, and nature ruthlessly imposes its own constraints? In a world facing many challenges as it prepares to enter the next century, this may be the most challenging of all.

Estimates of future numbers are based on the latest United Nations population projections, using their medium level figures. Under this scenario, world population will grow from 6.1 billion in 2000 to 9.4 billion in 2050—a gain of 3.3 billion. The other two U.N. projections put global population in 2050 as high as 11.2 billion or as low as 7.7 billion. While the medium scenario is judged by the U.N. demographers as the one most likely to materialize, it is not an inevitable population part for the next century. Indeed, because the projections are based exclusively on demographic assumptions and do not take into account the environmental limits to carrying capacity, they should be viewed as a first pass rather than the final word on estimates of future population.

We use the medium-level projections to give an idea of the strain this "most likely" outcome would place on ecosystems and governments, and the urgent need to break from the business-as-usual scenario. The mid-level projected growth in population of 3.3 billion by 2050 is very close to the growth that will have occurred between 1950 and 2000, some 3.6 billion. But there is one difference. During the half-century now ending, the growth occurred in both industrial and developing countries. During the next half-century, the entire burden of the projected increase of 3.3 billion will be in developing countries, many of which are hard-pressed to satisfy even existing demands on resources. In fact, the population of the industrial world is expected to decline slightly.

The annual rate of world population growth reached its historical high in 1964 at 2.2 per cent. Since then, it has been slowly declining, dropping to 1.4 per cent in 1998. Despite the falling rate of growth the number of people aged each year increased from 72 million in 1964 to the all-time peak of 87 million in 1990. Since then the annual addition has also declined, falling to 80 million in 1997, where it is projected to remain for the next two decades before starting to decline.

The population projections for individual countries vary more widely than at any time in history. At mid-century populations were growing everywhere, but today they have stabilized in some 32 countries, while they continue to expand in some countries at 3 per cent or more a year. Indeed, the world can be divided demographically into two camps: countries that have achieved population stability or are well on the way to doing so, and those that have not.

With the exception of Japan, all the nations in the first camp are in Europe. And all are industrial countries. The populations of some countries, including Russia, Japan, and Germany, are actually projected to decline some what over the next half-century. In addition to the 32 countries, containing 12 per cent of world population, that have stabilized their populations, in another 39 countries fertility has dropped to replacement level (roughly two children per couple) or below. Among the countries in this category are China and the United States the first and third largest countries, which together contain 26 per cent of the world's people.

Although fertility in these 39 countries has fallen below replacement level, their populations have not yet stabilized because there is a disproportionately large number of young people moving into the reproductive age group. Thus even if they hold their fertility at replacement level, population may continue to grow for several decades before it stabilizes. It was this realisation that led China nearly 20 years ago to shift its goal from a two-child to a one-child family. Leaders in Beijing realized that, if they did not do this they would be faced with adding the equivalent of another India to their population—a development they considered potentially disastrous for their people.

In contrast to this group some countries are projected to triple their populations over the next half-century. For example, Ethiopia's current population of 62 million will more than triple, as it climbs to 213 million in 2050. Pakistan's population is projected to go from 148 million to 357 million, surpassing that of the United States before 2050 today to 339 million, giving it more people in 2050 than there were in all of Africa in 1950. From an environmental Vantage point, considering particularly the availability of water and cropland, it is unlikely that the projected population increases for these three countries, and other countries with similar projected gains, will materialise.

As hard as it is to imagine the addition of another 3.3 billion people to the world's population, it is even more difficult to understand the effects of adding such numbers. As we look back over the last half-century, we see that World lumber use more than doubled, paper use increased nearly sixfold, grain consumption nearly tripled, water use tripled, and fossil fuel burning increased some fourfold. The relative contribution of population growth and rising affluence to the growth in demand for various resources varies widely. With lumber use, most of the doubled use is accounted for by population growth. With paper, in contrast, rising affluence is primarily responsible for the growth in use.

One way to understand the consequences of future population growth is to contrast some of the key trend projected for the next half-century with those of the as one. For example, we have seen a new fivefold growth in the oceanic fish catch and a doubling in the supply available per person, but biologists now believe we may have "hit the wall" in oceanic fisheries and that the oceans cannot sustain a catch any larger than today's. Thus people born today are likely to see the catch per person cut in half during their lifetimes.

Grainland per person has been shrinking since mid-century, but the drop projected for the next 50 years means the world will have less grainland per person than India has today. Future population growth is likely to reduce this key number in many societies to the point where they will no longer be enable to feed themselves. Countries such as Ethiopia, India, Nigeria, and Pakistan will see grainland per person shrink by 2050 to less than one tenth of a hectare (one fourth of an acre)—far smaller than a typical suburban building lot in the United States.

Given that at the amount of fresh water produced each year is essentially fixed by nature, the water available, per person has shrunk steadily as a result of population growth, leading to sever water shortages in some areas. Countries now experiencing these shortages include China and India, along with scores of smaller ones. As irrigation water is diverted to industrial and residential uses.

The challenge to governments presented by continuing rapid population growth is not limited to natural resources. It also includes education, housing, and jobs. During the last half-century the world has fallen further and further behind in creating jobs, leading to record levels of unemployment and under employment. Unfortunately over the next 50 years the number of entrants into the job market will be even greater. Few things threaten the political stability of a country as much as growing ranks of unemployed young people.

As noted earlier, the U.N. population projections cited here are based on exclusively demographic assumptions, which are not related to the population carrying capacity of local eco-systems. These projections are purely statistical, based on historical data on fertility, mortality, and average life span and assumptions about future trends.

Based on the analysis in it, I conclude that the medium projection of 9.4 billion people in 2050 which U.N. demographers consider to be the most problem is unlikely to materialize. Rather the world is more likely to follow a path closer to the low population projection of 7.7 billion by mid-century.

What is less clear is whether we will move to the lower trajectory because countries with rapid pollution growth quickly shift to smaller families or because they fail to do so and the resulting inability to manage threats from disease, spreading hunger, or social disintegration leads to rising death rates.

7

What is Known about Reducing Maternal Mortality?

Historical records demonstrate the significant improvements that can be achieved when key interventions are in place. Reductions in maternal mortality took place in Sweden during the 1800s, for example, as a result of a national policy favouring professional midwifery care for all births, coupled with establishment of standards for quality of care. By the beginning of the 20th century, maternal mortality in Sweden was the lowest around 230 per 1,00,000 live births compared with over 500 per 1,00,000 in the mid-1880s. In Denmark, Japan, Netherlands, and Norway, similar strategies produced comparable results. In England and Wales, significant reductions in maternal mortality were not apparent until the 1930s; at the national level, political commitment to the strategy was achieved only slowly and the introduction of professional midwifery was correspondingly delayed. In every case, however, the key to these improvements was the institution of fully professional maternity care.

In the USA, where strategy focused on hospital delivery by doctors, maternal mortality remained high because it proved difficult to establish adequate regulatory frameworks and mechanisms to ensure quality of care. In 1930, the maternal mortality ratio in the USA was still 700 per 1,00,000 live births compared with 430 in England and Wales.

More recently, India witnessed significant reductions in maternal mortality in a relatively short period. From a level of over 1500 per 1,00,000 live births in 1940-1945, maternal mortality fell to 555 per 1,00,000 in 1950-1955, 239 per 1,00,000 within 10 years, and 95 per 1,00,000 by 1980. The figure is now 30 per 1,00,000. These improvements followed the introduction of a system of health facilities around the country allied to an expansion of midwifery skills and the spread of family planning. During the 1950s most births in India took place at home with the assistance of untrained birth attendants. By the end of the 1980s over 85% of all birth were attended by trained personnel.

Similar evidence of the effectiveness of health care interventions is available from China, Cuba, and Malaysia. These countries established community-based maternal health care systems comprising prenatal, delivery, and postpartum care and a system of referral to a higher level of care in the event of obstetric complications.

What these examples clearly demonstrate is that a country's overall economic wealth is not in itself the most important determinant of maternal mortality. There are numerous other examples of countries with modest levels of GNP which have achieved low maternal mortality.

8

Safe Motherhood is a Human Rights Issue

The death of a woman during pregnancy or childbirth is not only a health issue but also a matter of social injustice. Of the human rights currently acknowledged in national constitutions and in regional and international human rights treaties, many can be applied to safe motherhood. Many such treaties and conventions are based on the 1948 Declaration of Human Rights; They include (1) the Convention on the Elimination of All Forms of Discrimination against Women, (2) the Convention on the Rights of the Child, (3) the European Convention for the Protection of Human Rights and Fundamental Freedoms, (4) the American Convention on Human Rights, and (5) the African Charter on Human and Peoples' Rights (6).

Human rights of relevance to safe motherhood can be grouped into the following four principal categories:

- ***Rights relating to life, liberty and security of the person,*** which require governments to ensure both access to appropriate healthcare during pregnancy and childbirth, and women's rights to decide whether, when, and how often to bear children. Governments must therefore address factors within the economic, legal, social and health systems that deny women these fundamental rights.
- ***Rights relating to the foundation of families and of family life,*** which require governments to provide access to health-services and other facilities that women

need to establish families and to enjoy life within a family.

- ***Rights relating to healthcare and the benefits of scientific progress, including health information and education,*** which require governments to provide access to good sexual and reproductive healthcare with appropriate referral systems. The measures needed to ensure safe motherhood can be provided through primary healthcare irrespective of a country's level of economic development. Central to these rights is information on a range of reproductive health issues, including family planning, abortion and sex education.
- ***Rights relating to equality and non-discrimination,*** which require governments to provide access to services such as education and healthcare without discriminatory grounds such as sex, marital status, age and socio-economic class. Discriminatory policies include requirements for a woman to obtain the consent of her husband for particular healthcare interventions, requirements for parental authorisation which have a differential impact on girls, and laws that criminalise medical procedures that only women need. Governments are in violation of their obligations when they fail to implement laws that effectively protect women's interests or to allocate health resources to meet women's particular need for safe pregnancy and childbirth.

The actions that governments need to take to promote safe motherhood as a human right fall into three groups:

- ***Reform of laws*** that prevent women from attaining the highest possible levels of health and nutrition needed for safe pregnancy and childbirth and that inhibit access to reproductive health information and services such as laws requiring women in need of health care to seek the authorisation of husbands or other family members first.

- ***Implementation of laws*** that foster women's right to good health and nutrition and that protect women's health interests such as laws that prohibit child marriage, female genital mutilation, rape and sexual abuse. Every effort should be made to implement laws that encourage the healthy timing of births, such as those that support the education of girls, set a minimum age for marriage and ensure women's access to essential health care.
- ***Application of human rights*** in national legislation and policy to advance safe motherhood.

9

Ecosystems, Our Unknown Protectors

How ecosystems work and what part they play in biodiversity remain a mystery. But we do know that they perform a host of invaluable services for the human species. In my view, biodiversity's fundamental value is neither aesthetic nor economic but environmental, even though most people are largely unaware of this. The value of biodiversity is often measured in terms of the number of species living in a given area. But the interactions between the many species in an ecosystem, and between them and the environment's physical and chemical components are also very important. This highly intricate web of relationships makes an ecosystem more valuable than the sum of the species it contains.

Ecosystems perform services that are essential for the survival of the human species. They fix carbon in the atmosphere and produce oxygen, protect soil from erosion and keep it fertile, filter water and replenish aquifers, provide pollination and anti-parasite agents and so on.

The first two of these services are closely related to each other. They result from photosynthesis, whereby green plants, starting with algae, absorb carbon dioxide (CO_2) and emit oxygen. For millions of years, the balance between the various gases in the atmosphere remained stable. But with the coming of the industrial revolution, humans began burning increase amounts of fossil fuels. Today, three billion tonnes of carbon build up in the atmosphere each year and natural ecosystems can no longer absorb all these emissions—especially since they are disappearing at an alarming rate. Deforestation alone releases such tremendous amounts of CO_2 and other gases, such as methane, that it has become the second-leading cause of global warming.

Storing freshwater, protecting soil and keeping it fertile are three other closely related functions. Ecosystems are veritable "freshwater factories". They absorb rainwater and slowly filter it through the soil before draining it towards streams, rivers, lakes and underground aquifers that supply us with the precious liquid. When the vegetal ground cover is degraded, the water cycle is disrupted. Rain strikes the bare earth, washing away huge amounts of nutritional substances. Reservoirs, lakes and rivers silt up.

Uncertain Reaction to Climate Change

Despite years of research, scientists still know very little about how ecosystems work. We are generally incapable of predicting how they will react to certain transformations in the environment, especially climate changes. Nor do we know any more about whether a species present in a given environment is superfluous or "replicable", even when it is very rare. Likewise, we do not know which key species are indispensable to maintaining an ecosystem, with a few exceptions such as pine forests, where that tree is obviously the dominant species.

We know even less about the part biological diversity itself play in maintaining ecosystems and the services they perform. One simple example is a highest diversified forest that absorbs carbon dioxide—a vital function, as we have seen, for limiting global warming. Suppose the forest is cleared to make way for a single-crop forest. The service will still be performed, perhaps even better at first because young, fast-growing trees absorb more CO_2 than old ones, which regenerate slowly. But what will happen in the long term? After several decades, the consequences of the loss of biodiversity will probably be felt. Replacing many species with a single one will have certainly depleted the soil and, in the long term, slowed down the forest's growth and consequently its ability to absorb CO_2.

More generally, diversified ecosystems seem more productive. Specialists remain wary about their conclusions, but today they believe that biodiversity helps ecosystems to resist alien species and diseases and to recover faster in the event of disruption. In the face of doubt, and to find out more about them, it is better to preserve as many different ecosystems as possible.

A Costly Lesson for New York City

Most people take it for granted that ecosystems will carry on performing services without receiving anything in return. They think nature will continue benefiting humanity, no matter how much damage is done. The survival of organisms other than our own species is perceived as a frill that future generations can live without.

These preconceived ideas are wrong and dangerous as the city of New York has recently come to realize. The city's water has always enjoyed such a good reputation that it was sold throughout the northeastern United States. Its equality was due to Catskill Mountains' natural purification system. But that ecosystem has suffered so much from pollution, especially fertilizer run-off from farms, that by the late 1990s New York's water had become undrinkable. The city planned to build a purification plant, whose cost was put at between six and eight billion dollars, not including the $300 million in yearly operating costs—an astronomical bill for a service that had always been free! The price was so staggering that the city eventually decided to restore the Catskill Mountains' degraded environment at a cost of only one billion dollars.

This story clearly illustrates where our interests lie. We must preserve ecosystems and the conditions that enable our planet to ensure the survival of *Homo sapiens* or, at least, the short-term maintenance of our current quality of life.

10

Forests

The Earth's Lungs

The world's forest cover is shrinking. Over the past 50 years nearly half of the world's original forest cover has been lost—some 3 billion hectares. Each year another 16 million hectares of virgin forest are cut, bulldozed, or burned.

Between 1980 and 1995 the world lost some 180 million hectares of forest—an area the size of Indonesia. While developed countries had a net increase of 20 million hectares due to reforestation, this gain was more than offset by a net decrease of 200 million hectares in the developing world.

Forests have many functions of value both to humanity and to nature itself. Take away the trees, and the intricately linked ecosystem unravels. Forests absorb carbon dioxide and produce oxygen, anchor soils, regulate the water cycle, protect against erosion, and provide a habitat for millions of species.

Forest products are essential to the world economy, worth about US$400 billion annually in timber, pulp, paper, and fuel wood. Forest products other than wood, such as medicines, vegetables, and fruits, provide another US$20 billion and are growing in importance.

Healthy forests boost food production. Trees soak up and store water from season to season, slowly releasing moisture during dry periods. Without tree cover, water runs off faster during the tropical rainy season, carrying away valuable topsoil. A World Bank study found that the rate of soil loss was 10 times higher on forest lands where slash-and-burn shifting cultivation was practiced than in undisturbed forests.

One reason that agricultural yields have fallen in sub Sahran Africa is that vast amounts of forests cover have disappeared, hastening soil erosion and loss of soil nutrients.

Forest cover regulates climate, while destruction of forests contribute to global warming. Whereas living trees soak up and store carbon dioxide from the atmosphere trees that are cut down and burned release carbon into the atmosphere. In the last decade tropical deforestation has released large amounts of stored carbon—accounting for roughly one-quarter of the carbon dioxide emissions to the atmosphere due to human activity.

Pressures on Forests

Current demand for forests products may exceed the limits of sustainable consumption by 25 per cent. The developed world accounts for most of the demand for forest products. With just 16 per cent of the world's population, North America, Europe, and Japan consume two-thirds of the world's paper and paperboard and half its industrial wood. Demand for industrial wood products also has risen in developing countries, however, along with demand for fuel wood, the main energy source for many rural communities.

Throughout the 1990s many developing countries with rapid population growth had high rates of deforestation. Forest land was converted to agricultural use, and trees cut to provide housing and wood for fuel. Moreover developing countries stepped up exports of forests, products to meet the rising demand from developed countries.

The amount of forest area per capita fell by half between 1960 and 1995—reflecting both population growth and the disappearance of forests cover. In 1995 close to 1.7

billion people lived in countries with less than one-tenth of a hectare of forest cover per capita (83). By 2025, an estimated 4.6 billion people will live in such countries.

What Can Be Done?

As population grows and per capita consumption of forest products increases, countries must do more to manage forest resources on a sustainable basis. The following developments offer encouragement.

Technological Improvements

Technological improvements including use of recycled paper and paperboard, have substantially reduced the amount of pulp needed to produce paper. In 1970 paper and paperboard consisted of 80 per cent wood pulp. By 1997 more efficient production processes had reduced that figure to 56 per cent. As a direct result, the production of pulp for paper is expected to grow by just over 1 per cent a year over the next decade, about half the growth rate in the 1980s.

Forest Products Certification

Adopting a system that identifies forest products that come from sustainable managed forests could support efforts toward sustainability. As of 1998, about 10 million hectares of forest lands have been certified. Over 90 per cent of the certified area is in northern, temperate forests, mostly in Europe and North America. Close to 60 per cent of the entire certified area is in just two countries—Sweden and Poland—reflecting education and awareness campaigns in those countries. In tropical forests, where most of the destruction is taking place today, only tiny areas have been certified as providing sustainable yield.

Intergovernmental Responses

In 1995 the Intergovernmental Panel on Forests (IPF) was established in response to the 1992 Earth Summit. The IPF evolved into the inter governmental Forum on Forests in 1997, after the UN's five year review of the Earth Summit goals. The mission of the forum is to examine the underlying causes of deforestation and to help countries develop strategies that address them.

Efforts to advance an international legal convention on forests, which begin in 1990, have been shelved, however. Some observers believe that advancing such a convention would only codify the standards of a weak consensus and thus would be worse than no convention at all. Widespread opposition to a convention makes it unlikely that the issues will reach the negotiating table.

Instead, many organisations urge governments of countries with large forest resources to enforce existing legislation and to introduce more effective forest conservation initiatives close to 130 countries have developed or updated their National Forest Programmes over the past decade.

While such initiatives are promising, they cannot be expected to half forest destruction completely. Millions of people rely on forest products for their livelihoods. Sustainable forest management will require not just enforcement of laws that project forests but also alternative sources of livelihood for many rural people.

11

Biodiversity

As human population has surged this century, the populations of numerous other species have tumbled, many to the point of extinction. Indeed, we live amid the greatest extinction of plant and animal life since the dinosaurs disappeared some 65 million years ago, with species losses at 100 to 1,000 times the natural rate. But humans are not just witnesses to a rare historic event, we are actually its cause. The leading sources of today's species loss, habitat alteration, invasions by exotic species, pollution, and overhunting are all a function of human activities.

Human activities have pushed the percentage of mammals, amphibians, land fish that are in "immediate danger" of extinction into double digits. The principal cause of species extinction is habitat loss—the result of encroachment by humans for settlements, for agriculture, or to claim resources such as timber. A particularly productive but vulnerable habitat is found in coastal areas, home to 60 per cent of the world's population. Coastal wetlands nurture two-thirds of all commercially caught fish, for example. And coral reefs have the second highest concentration of biodiversity in the world, after tropical rainforests. But human encroachment and pollution are degrading these areas: roughly half of the world's salt marshes and mangrove swamps have been eliminated or radically altered, and two thirds of the world's coral reefs have been degraded, 10 per cent of them "beyond recognition". As coastal migration continues—coastal dwellers could account for 75 per cent of world population within 30 years—the pressures on these productive habitats will likely increase.

Habitat loss tends to accelerate with an increase in a country's population density. This is bad news for the world's biodiversity hotspots-species-rich ecosystems at greatest risk of destruction. Twenty-four of these hotspots, containing half of the planet's species, have been identified globally. Some of the most important hotspot countries will reach population densities that have been linked with very high rates of habitat loss. Five of the six most biologically rich countries could see more than two-thirds of their original habitat destroyed by 2050 if this historical relationship holds.

Related to loss of habitat is the growing incidence of plant, animal, insect, and microbial invasions of ecosystems worldwide as human interchange increases. These "exotic species" sometimes dominate local ecosystems, eliminating native species and reducing overall diversity. Exotics are implicated in 68 per cent of all fish extinctions in the United States this century, for example. Growth in human travel and commerce explains many accidental invasions by exotics, but foreign species are also deliberately introduced into farms, plantation forests, and aquaculture systems. Although only 1 per cent of exotics cause widespread damage, exotic species are the second leading cause, after habitat destruction, of species loss worldwide.

Other, often diffuse effects of expanded human activities also disrupt ecosystems. Nitrogen, for example, is now made available to plants at more than twice the preindustrial rate as a result of fertilizer production, cultivation of nitrogen-fixing crops, and the burning of fossil fuels. This overfertilisation of the Earth favours some species at the expense of others, leading to a reduction in diversity and resiliency of land and aquatic ecosystems.

Likewise, greenhouse gas emissions could disrupt ecosystems on a vast scale. As with nitrogen, increased levels of atmospheric carbon may favour some species over others:

annuals over perennials, for example, or deciduous trees over evergreens. To the extent that greenhouse gases induce changes in global climate, many species may be at risk as habitats shift or shrink, and as some life forms, such as insects or animals, adapt and migrate more quickly than others, such as plants. And as sea levels rise with a change in climate, ecosystems such as coastal wetlands could be destroyed.

12

Living with Diversity

Fishers' nets and loggers' saws may directly impoverish local ecosystems, but most biological losses have root causes far away, in long-settled urban areas and farms where diversity is seldom a concern, but where steadily rising demand for food, water, wood and other resource—and the dispersal of resulting wastes—reach far beyond the settled areas themselves. In general, these peopled landscapes have lost much of their own biological wealth, but what remains is still important to their continued functioning and livability. Reconciling farms and cities with diversity will require stopping the damage they bring to remaining natural habitats, but also beginning to halt and reverse the homogenisation of these unnatural habitats.

Uniformity is not inherently undesirable. In fact, to some degree, homogeneity is the basis of all agriculture: a given type of plant is favoured and others are suppressed or eliminated. But trends in recent decades (most notably the Green Revolution and the parallel intensification of farming systems in industrial nations) have pushed uniformity to dangerous levels.

The unsustainability of modern agriculture is in part a measure of its inability to tolerate diversity. Both genetic and ecological uniformity—the sameness of fields sown horizon to horizon without interruption—demand costly and often futile reliance on chemicals to protect crops from pests or diseases that are rapidly spreading and evolving. The drive to leave no hectare unplowed worsens soil erosion, pushing tractors onto highly erodible hillsides and removing windbreaks, hedgerows and other remnant habitats.

Some of agriculture's biological impacts are obvious—the expansion of farms onto forests and wetlands, for example. While the increasing reliance on chemical inputs and machinery has reduced these impacts in some cases by decreasing the area needed to produce a given amount of food, it has worsened others.

A fundamental transition away from today's wasteful and polluting farming systems is needed to put the world's food supplies on a secure footing. Many of the reforms that will reduce farming's dependence on fossil fuel inputs and its misuse of soils and waters can also restore diversity to agricultural landscapes. Pesticides, for example, kill not only pests but other animals, such as pollinators and predators, that are beneficial to agriculture. Alternative pest control measures that lower pesticide use can also, ironically, reduce pest damage to crops by reviving the diversity of soil and insect communities, which play crucial roles in maintaining soil productivity and checking the spread of pest outbreaks.

Traditional agroecosystems are important not only because they provide sustenance to rural people and harbour valuable genetic resources, but also because they contain the seeds of a sustainable, diversity-based mode of agriculture. At varying levels, diversity is the basis of production for many peasants. Farmers often mix strains of a given crop in their fields as a hedge against the vagaries of weather. They also tend to recognize the dependence of their farms on adjacent ecological systems and to tolerate wild plants (often crop relatives whose continued interbreeding with domestic descendants contributes to genetic variety) on the outskirts of their fields.

Population growth and the expansion of large commercial farms have rendered many once-sound practices no loner viable, and traditional agriculture badly needs usions of money and research to increase its modest yields without abandoning its stability.

Urban areas, with good reason, are considered the antithesis of natural diversity. Only the most resilient creatures (many of them regarded as weeds and pests) thrive in them, and cities' ceaseless expansion, consumption of resources, and emissions of waste threaten both farmland and wilderness almost everywhere. As with agricultural lands, the first priority for urban areas is to half their expansion onto other ecosystems and reduce the damage they export, such as the sewage poured onto coral reefs by burgeoning coastal cities throughout the tropics, or the wasteful consumption of tropical hardwoods in Japanese building construction.

But even concrete jungles can support some diversity. Landscaping of private yards and public spaces with native vegetation can not only reduce the expense and environmental impact of watering, spraying and hauling the remains of sterile grass monocultures, but also help revive bird and other wildlife populations. Most urban areas also have water-ways running through them, or corridors of unused land such as steep ravines; if their use as waste receptacles is reduced, these can be maintained or restored as wildlife habitat.

In developing nations, especially, a surprising amount of agricultural production takes place within city limits, in home gardens. These hidden farmlands contain a great deal of genetic diversity, and their expansion could help reduce the scale and environmental impacts of commercial agriculture.

One reason that the destruction of biological diversity has gone so far without major public commitments to stopping it is that urban dwellers have little experience of the natural and even less understanding of its importance. Restoring nature where people live—reestablishing a personal link with the living world—may be necessary to save it elsewhere. For all the rational arguments favouring long-term protection of biological assets, people who have lost all direct sense of their dependence on natural systems may simply not care.

Only a growing respect for diversity for its own sake—beginning, perhaps, with a reconnection between people and nature within the urban environment—will trigger altruistic responses among those wealthy enough to have the option of considering the needs of future generations and natural communities. Although many conservation measures make economic sense, arguments of economics or self-interest will likely fail to be convincing when the contest is between a few uncharismatic species of unknown value and a major industrial project. "Human beings make sacrifices for what they love." Those who maintain strong bonds with the biological world on which they depend may be more inclined to make the hard decisions needed to protect it.

13

Global Warming

Worrisome Signs

Scientists increasingly agree that the earth's atmosphere is becoming warmer. A long-term rise in the global climate could cause sea levels to rise around the world and bring a number of other adverse consequences. Reliance on fossil fuels as an energy source and the widespread destruction and burning of forests are chiefly responsible for the carbon emissions—the so-called greenhouse gases—that lie behind global warming.

One indication of global warming is that over the past 40 years the ocean surface (the top 1,000 feet) has warmed an average of half a degree Celsius. The US National Oceanic and Atmospheric Administration (NOAA) has reported that tropical waters in the Northern Hemisphere have been warming up even faster—in fact, 10 times faster than the measured global rate—because tropical oceans retain heat more readily than other areas.

Rising Sea Levels

Studies project that by 2100 the earth's surface temperature could increase between 1.0 and 3.5 degrees Celsius. If the highest projection were reached, Greenland's ice sheet probably would melt. As a consequence, the global sea level gradually would rise as much as seven metres.

Computer models project that this rise in sea level would take more than a millennium. Some climatologists, however, think that sea levels could rise much faster, pointing to dramatic shrinkage of the Arctic ice cap over the past 30 years.

Even a rise of one metre in sea level which could occur by 2080, according to the computer models would inundate many low-lying coastal areas around the world. For instance, much of the Nile River Delta of Egypt would disappear. A one-metre rise in global sea levels also would inundate close to 20 per cent of the coastline of Bangladesh and displace millions of people.

Adverse Health Effects

Rising global temperatures also would carry adverse health consequences. As temperatures warmed and episodes of droughts and floods became more frequent, the incidence of water-borne diseases and a resurgence and spread of infectious diseases carried by mosquitoes and other disease vectors probably would increase.

Warmer global temperatures also would magnify the effects of human activities on the environment, including more pollution and habitat destruction. Climate change might even cause some ecosystems to exceed critical thresholds, leading to their irreversible decline.

Growing Scientific Consensus

In 1988, to help study and focus attention on the issues, the Intergovernmental panel on climate change was created under the auspices of the World Meteorological Organisation and the United Nations Environment Programme. The panel has involved as many as 2000 scientists from around the world. In 1996 a panel report concluded firmly that global climate change is a reality and not just a possibility. After reviewing the evidence, the panel determined that:

- Evidence for the link between climate change and human activities is compelling. Already, increases in carbon dioxide and other climate changing gases have upset the balance of the earth and its atmosphere.

- The earth's surface has become warmer; the number and severity of storms have increased; and the global sea level has risen by 10-25 cm. over the past century.

Because the warming trend is a global problem, solutions must be global in scope, the panel concluded.

Why is the Climate Changing?

Over the last 150 years burning of fossil fuels has released some 270 billion tons of carbon into the atmosphere in the form of heat-trapping carbon dioxide gases. Since 1950 annual worldwide carbon emission have increased fourfold, reaching 6.3 billion tons in 1997. Other emissions that contribute to climate change include methane (mainly from domestic livestock and agriculture), nitrous oxide, and chlorofluorocarbons.

Atmospheric concentrations of carbon dioxide reached 363 parts per million in 1998, the highest level since the time of massive volcanic activity over 160,000 years ago, based on examination of ice cores in Antarctica and in the Arctic. If current trends continue, atmospheric concentrations of carbon dioxide would double during this century.

About three-fourths of the huge increase in carbon emissions over the past half-century is due to increased energy consumption per capita; about one-quarter is due to population growth. Western industrialized countries account for nearly half of atmospheric carbon emissions, but developing countries are producing a growing share as industrial activity increases and populations grow. China is now the world's second largest carbon emitter, after the US.

Vanishing Carbon Sinks

The earth's forests are carbon sinks that currently soak up an estimated one-third of the carbon dioxide released into the atmosphere. When forests burn, whether naturally or when people clear the land, they not only release more carbon into the atmosphere but also diminish the amount of carbon-absorbing forest cover remaining.

Some scientists are concerned that droughts caused by global warming will increase the number of forest fires, thus contributing further to carbon emissions in the atmosphere. For instance, the six months of extensive forest fires that occurred in Asia in 1997 and 1998 released more carbon into the atmosphere than western Europe emits in a year. Burning trees for land clearance in the tropics releases about 1 billion tons of carbon into the atmosphere annually.

As more carbon fills the atmosphere, scientists worry that forests will become saturated and no longer play their role as carbon sinks. Instead, they will start to release carbon themselves.

Agriculture at Risk

Higher carbon dioxide levels in the atmosphere would extend the agricultural growing season and promote forest growth in the short run but would have potentially negative effects on crops and forests in the long run. Because the world's grain belts would become less productive, an additional 350 million people would go hungry by the middle of this century. Major droughts have been projected for sub-Saharan Africa as climatic patterns shift, reducing rainfall and drying out soils for longer periods.

In 1999 NOAA project that by the middle of this century soils in agricultural regions of the Central US, Central Asia, and the areas surrounding the Mediterranean Sea would likely experience substantial reductions in soil moisture during the summer growing season because evaporation rates would be higher. Such reductions in soil moisture would make these areas particularly vulnerable.

Others point out that, ironically, global warming could produce colder temperatures in Northern Europe and Russia, reducing crop yields in these regions as well. This change would occur because the huge amounts of arctic fresh water from melting ice caps would make the water less dense. Such a change would interrupt the "conveyor belt" effect of the North Atlantic Drift, the ocean current that transports warm tropical water from the Gulf Stream to Scandinavia and Northern Europe.

What Can be Done?

What is the prospect for reducing emissions of carbon dioxide into the atmosphere? The United Nations Framework Convention on Climate change was opened for signature at the Rio Earth Summit in 1992. It was promptly signed and ratified by most low-lying island states and countries with extensive coastal areas. The Convention established a framework and a process for agreeing on specific actions later on; it asked signatory states to take preliminary action to reduce greenhouse gas emissions; and it encouraged scientific research on climate change.

14

Climate Change

Over the last half-century, carbon emissions from fossil fuel burning expanded at nearly twice the rate of population, boosting atmospheric concentrations of carbon dioxide, the principal greenhouse gas, by 30 per cent over preindustrial levels. All major scientific bodies acknowledge the likelihood that climate change due to the buildup of greenhouse gases in the atmosphere is indeed under way. The 15 warmest years on record have all occurred since 1976, and 1998.

The destabilisation of our climate threatens more intense heat waves, more severe droughts and floods, more destructive storms, and more extensive forest fires. The related shifts in rainfall and temperature may jeopardize food production, the Earth's biological diversity, and entire ecosystems, as well as human health by expanding the ranges of tropical diseases. Unless efforts to curb them are stepped up, carbon emissions will continue to grow faster than population over the next 50 years, driving the Earth's climate system into unchartered territory. The Intergovernmental Panel on Climate Change (IPCC) estimates that an eventual two-thirds reduction in global emissions is needed to avoid precariously high levels of atmosphere carbon dioxide concentrations.

The IPCC and U.S. Department of Energy (DOE) project that emissions from developing countries will nearly quadruple over the next half-century, while those from industrial nations will increase by 30 per cent. Although annual emissions from industrial countries are currently twice as high as from developing ones, the latter are on target to eclipse the industrial world by 2020.

Higher per capita carbon emissions accounts for roughly 55 per cent of the increase in emissions projected for developing nations. Emissions per person are due to more than double from 0.51 tons of carbon per year in 2000—just one-fifth of the industrial level—to 1.14 tons in 2050. The remaining 45 per cent of emissions increases is due to population growth.

Fossil fuel use accounts for roughly three quarters of world carbon emissions. As a result, regional growth in carbon emissions tend to occur where economic activity, and related energy use, is projected to grow most rapidly. Emissions in China are projected to grow over three times faster than population in the next half-century, as emissions per person soar from 0.77 tons of carbon to 2.81 tons due to booming economy that is heavily reliant on coal and other carbon-rich energy sources. In Africa, in contrast, emissions per person are expected to scarcely change-growing from the current level of 0.33 tons in 2050, despite a threefold increase in total emissions.

The effects of population growth are most profound in countries where people are heavily emitters. For example, the 115 million people added to the population of the United States between 1950 and 1998—an increase of nearly 75 per cent in just 45 years—account for more than one-tenth of current global emissions. And the carbon emissions of the 75 million people who will be added to the U.S. population in the next 50 years roughly equal the emissions of the 1.3 billion people who will be added to Africa during that period.

Deforestation and other land use changes account for the remainder of world carbon emissions. Forests have served as a sink for carbon throughout much of human history. In recent years, however, the world's forests have become net sources of atmospheric carbon, largely due to forest burning and clearing in the tropics. Six months of fires in Asia in

1997 and 1998 released more carbon than Western Europe emits from fossil fuel burning in an entire year. The carbon contribution from this source will likely increase in coming years as the burgeoning human population continues to cut down forests.

15

Forests

Global losses of forest area have marched in step with population growth for much of human history. The two trends rose slowly for millennia, turned upward in recent centuries, and accelerated sharply after 1900. Indeed, 75 per cent of the historical growth in global population and an estimated 75 per cent of the loss in global forested area have occurred in the twentieth century. The correlation makes sense, given the additional need for farmland, pastureland, and forest products as human numbers expand. But since 1950, the advent of mass consumption of forest products has quickened the pace of deforestation.

In some cases, population pressure is still closely linked with deforestation. In Latin America, for example, ranching is the single largest cause of deforestation. Because most meat produced in Latin America is consumed there, and because meat consumption per person has been largely unchanged for several decades, it is likely that expanding population is the principal reason for ranching-related deforestation. In addition, analysts at the World Resources Institute estimate that overgrazing and overcollection of firewood which are often a function of a growing population are degrading some 14 per cent of the world's threatened frontier forests (large areas of virgin forest). In fact, a U.N. Food and Agriculture Organisation study showed a one-to-one correlation between population growth and fuelwood consumption in 16 Asian countries between 1961 and 1994.

On the other hand, deforestation created by the demand for forest products tracks more closely with rising per capita consumption in recent decades. Global use of paper and paperboard per person, for example, has doubled (or nearly tripled) since 1961, and most of the increase has come in wealthy countries with low or even stable levels of population growth. Europe, Japan, and North America, with 16 per cent of global population, consume 63 per cent of the world's paper and paperboard and nearly half its industrial wood.

Although consumption and population growth have operated somewhat independently in the late twentieth century, the two forces could coincide in the developing world in coming decades, with substantial consequences for forests. Developing country paper consumption is less than one-tenth the level found in industrial nations, suggesting that large increases in consumption are likely as these nations prosper. (It also suggests that greater economy is needed in industrial countries). With 80 per cent of the world's people, and as home to all the increase in population in coming decades, even modest growth in per capita paper and wood consumption in developing countries could place substantial pressure on forests. If paper were used by the entire world in 2050 at today's industrial-nation rates, paper production would need to jump more than eightfold over 1996 levels.

This projected growth is unsustainable, given that global use of forest products is already near or beyond the limits of sustainable use. Using data on sustainable forest yields, and assuming that virgin forests are left intact, researchers at Friends of the Earth UK have determined that production of forest products for the world is 25 per cent beyond the most restrictive estimates for sustainable consumption. (Many forests, of course, are already logged well beyond sustainable levels). The most optimistic assessment would allow for a further 35 per cent growth in consumption. Even that spells trouble, however, given a projected global population increase of some 54 per cent over the next half-century, and given the likely increase in consumption from rising prosperity. Lower consumption of forest products and increased recycling in industrial countries can make room for a more prosperous developing world to enjoy the products of the world's forests, but the task will be made easier if population growth everywhere is stabilized sooner rather than later.

If population and consumption eat into the world's forests, the resulting loss of forest services reduces, in turn, a country's capacity to support its population. Forests provide habitat to a diverse section of wildlife; tropical forests, for example, are home to more than 50 per cent of the world's species. And as storehouses of carbon, forests are key to regulating climate. Deforestation leads to huge releases of carbon: an estimated one-quarter of the world's carbon emissions come from forest clearing. Loss of these macroservices undermines the stability and resiliency of the global environment on which economies—and populations—depend. In addition, forests provide services vital to a local population, such as control of erosion, steady provision of water across rainy and dry seasons, and regulation of rainfall. Taken together, the loss of these services due to deforestation can upset local economies and subject local populations to economic instability.

16

An Agenda for Change

The world's growing population, combined with unsustainable production and consumption patterns, is putting increasing stress on air, land, water, energy, and other essential resources.

- Development strategies will have to deal with the combination of population growth ecosystem health, technology, and access to resources. Meeting the unmet need for family planning and reproductive health services should be part of national sustainable development strategies.

- The world needs to do a better job of forecasting the possible outcome of current human activities, including population trends, per capita resource use, and wealth distribution.

Protecting the Atmosphere. The atmosphere is under increasing pressure from green house gases that threaten to change the climate and from chemicals that reduce the ozone layer. Governments need to:

- Modernize existing power system to gain energy efficiency and develop new and renewable energy sources.

- Promote national energy efficiency and emission standards and develop efficient, cost-effective, and less polluting mass transit systems.

Combating Deforestation. Forests world wide are threatened by uncontrolled degradation and conversion to other uses because of increasing human pressure.

- There is an urgent need to conserve and plant forests in developed and developing countries to maintain or restore the ecological balance and to provide for human needs.
- Governments need to work with business, scientists, local community groups, indigenous people, and the public to create long-term conservation and management policies for every forest region and watershed.

Sustainable Agriculture and Rural Development. Hunger is already a constant threat to over 800 million people, while the world's ability to continue meeting growing demand for food and other agricultural products over the long term is uncertain. Soil erosion, salinisation, water-logging, and loss of soil fertility are increasing in all countries.

Agriculture has to meet rising needs mainly by increasing productivity, because most of the world's best croplands are already in use. At the same time further encroachment on land that is only marginally suitable for cultivation must be avoided.

- Sustainable agriculture and rural development will require major adjustments in agricultural, environmental, and economic policies in all countries and at the international level.

Conservation of Biological Diversity. The loss of the world's biological diversity continues, mainly from habitat destruction, over-harvesting, pollution, of foreign plants and animals (known and exotics). This decline in biodiversity is largely caused by human activity and represents a serious threat to our development.

- Develop national strategies to conserve and sustainably use biological diversity and to make these strategies part of overall national development efforts.
- Implement fair sharing of the benefits between providers and consumers of biological resources.
- Protect natural habitats. Promote the rehabilitation of damaged ecosystems.

Protecting and Managing the Oceans. Oceans are under increasing environmental stress from pollution over-fishing, and degradation of coastlines and coral reefs. About 70 per cent of marine pollution comes from sources on land. Countries should commit themselves to control and reduce degradation of the marine environment. They should:

- Build and maintain sewage treatment systems and avoid discharging sewage near shell fisheries, water intakes and bathing areas.
- Develop land-use practices that reduce run-off of soil and wastes to rivers and thus to the seas. Use environmentally less harmful pesticides and fertilizers.
- Control and prevent coastal erosion and silting due to land uses such as unplanned construction.

Protecting and Managing Fresh Water. In many parts of the world there is widespread scarcity, gradual destruction, and increased pollution of fresh water resources. The causes include the inadequately treated sewage and industrial waste, loss of natural water catchment areas, deforestation and other chemicals into the water. The following approaches are key:

- The way to provide all people with potable water and basic sanitation is to adopt the approach "some for all rather than more for some." This approach can be achieved through low-cost services built and maintained at the community level.

- Nations need to identify and protect water resources and see that water is used on a sustainable basis. They need effective water pollution prevention and control programmes. There is a particular need for appropriate sanitation and waste-disposal technologies for low-income, high-density cities.

17

Ecotourism or Ecocide?

Ecotourism is the fastest growing part of the World travel business, but whether it destroys more than it protects will depend upon how it is put into practice.

For the travel and tourism industry, ecotourism is the fastest growing 'market segment', generally equated with nature tourism. Interpreted merely as a product, however, it may be ecologically based but not ecologically sound, responsible for sustainable.

To incorporate these vital characteristics ecotourism must adhere to three essential principles: The first is, perhaps, the most obvious. As an industry based on the beauty and diversity of nature, it is evident that it should not deplete or degrade those resources and thus prejudice its own future. Ecotourism must, therefore, be ecologically sound, requiring a two-way link between itself and environmental conservation.

To consider nature without recognising the link with people will, however, compromise sustainability. It is now widely recognised that conservation cannot be divorced from development issues. The second principle is therefore that ecotourism must be responsible, paying regard to local needs and improving local welfare.

However, to be truly sustainable, ecotourism needs to fulfill the ambitions and expectations of all interests. The third principle, then, is to consider not only the interests of tourism enterprises and organisations, but also visitor satisfaction, the needs of tourists.

If, ecotourism embodies these essential principles, symbiotic relationships between the varying interests should follow, with environmental protection resulting both from and in enhanced standards of living for local populations, continued profits for the tourism industry, sustained visitor attraction, and revenue for conservation. An examination of ecotourism across these dimensions, highlights not only its potential but also its problems.

Local Benefits

The high ground claimed by ecotourism, in terms of its contribution to development, is that, in principle, it offers enhanced prospects for local involvement compared with conventional tourism. As well a moral obligation to incorporate the local people in projects that affect them, such incorporation has important developmental implications.

Tourism income may be captured locally through revenue sharing schemes, through entrepreneurship and labour, and through the sale of tourist merchandise. Tourism can also act as a catalyst, and even provide some of the finance, for the improvement of essential services such as clean water, sanitation, electricity supply and transport. It may also provide an incentive for improved education and skills and the potential for participation in decision making.

Local involvement also makes sense for conserving natural environments. It has been recognised that, as people realise the benefits from ecotourism, support for conservation increases. Ecotourism may also provide the incentive for the survival of a traditional culture. The cultural and the natural are often inextricably linked to form the composite attraction of a particular ecotourism destination. The terracing of the Himalayan foothills, the hot springs at Tatopani, Sikkim are all examples of the fusion of the natural and the cultural.

Greater local involvement makes practical sense for national and local governments, agencies and operators using local labour, expertise and knowledge. Education is a two-way process, improved understanding of local circumstances is likely to increase project efficiency. Building upon local experience and traditions provides a foundation for wise and successful development, and, simultaneously, an ecotourism asset.

Introduction to indigenous uses of natural products also broadens the base of environmental interpretation. Local involvement is not without its problems, however. Revenue sharing schemes may neither benefit the most needy, nor those most adversely affected. Beneficiaries may often be passive recipients, rather than active participants. The emphasis must be on participation rather than patronisation, if traditional livelihood are removed they must be replaced with others.

Local participation, however, often consists of employment rather than entrepreneurship, where constraints of costs of entry, language, education and skills operate. Furthermore, the nature of local employment tends to be low skilled, poorly paid and often seasonal. The higher status, better paid jobs, particularly managerial positions, tend to be occupied by outsiders.

Industry Profits

The integrity of the tourism 'product' is vital to the interests of tourism entrepreneurs, and all those who are associated with them, such as tourist boards, government departments, NGO's and international aid agencies. Tourism operators benefit from public support, increased credibility and demand for associated products. Sound environmental practice often makes good business sense.

There are, however, many practical and institutional obstacles to effective ecotourism management, not the least of which will be the problems of vested interests who are more concerned with short term profits than with the long-term.

Pressure of Numbers

Another dilemma is the sheer problem of numbers. To confine attention, however, to the consideration of small-scale, more easily managed ecotourism projects, involving small specialist groups paying high prices, is to invite ecocide at higher levels.

Rapid growth rates imply inevitable change. Psychological carrying capacity (as well as other types of carrying capacity) will probably be breached and visitor satisfaction compromised. This is especially true when visitors are concentrated in space and time.

However much a principled definition of ecotourism is advocated, it must be recognised that so-called ecotourists are not an homogeneous group. The spectrum of participants embraces hard-core nature tourists through to casual day visitors. Their behaviour and consequent impact will vary accordingly. It is essential, therefore, to attempt to match numbers and types of ecotourists with destination characteristics.

Paying for Conservation

Willingness-to-pay surveys of ecotourists across the globe show a consistent response of $10 as reasonable visitor's fee. Certain unique sites, or those harbouring more charismatic species, can support higher fees. The potential revenue for conservation is therefore evident, but often not realised.

Where the fee falls below the amount visitors are willing to pay, the capability to contribute more fully towards conservation remains latent. It is also necessary to ensure that a proportion of revenues accrues locally. Percentages of revenues directed towards conservation vary between sites. Often high proportions end up in central treasuries.

The Challenge

The major role players in ecotourism all have a stake in its sustainable development. Their present and future interests are, in many ways, tied to one another. Given the multitude, and diversity, of stakeholders a completely sustainable outcome is, however, likely to remain elusive.

The grand challenge is to reconcile sometimes complementary, but often conflicting interests. The essential dilemma is to balance demands of ever-increasing 'new-tourists' escaping from the confines and pressures of urban life, and reacting against the characteristics of mass tourism with the needs of the environment, the aspirations of tourism organisations, and, most importantly, the basic needs of the local population.

Although a win-win scenario, where all interests gain, is the ideal outcome, there will often be situations where one interest may gain at the expense of another. National Parks, for example, may bring benefits for conservation and for visitors, but the local population is likely to lose out if they are excluded from their traditional activities.

The situation is fraught with discontinuities. A win situation for one interest in a particular place at a specific point in time is likely to be a loss for another. It is necessary, therefore, to recognise conflicts and identify relative costs and benefits. Arriving at the most sustainable outcome is likely to involve trade-offs. It is unlikely to be optimal either environmentally or developmentally, but, in the circumstances, it will be the most feasible and most practical. And, hopefully, ecocide will be circumvented.

18

Urbanisation and the Environment

Is abandoning the cities the answer to the growing ecological problems of urbanisation? The trend at any rate is in the opposite direction. At the beginning of this century, only every 10th person worldwide was a city dweller. At its end, more than half the global population will be urbanites. And most of the urban population growth will take place in the developing countries, led by Asia.

Compared to other parts of the world, however, the urbanisation process in Asia is currently not even particularly far out in front. Worldwide, city dwellers account for 43 per cent of the total population. Industrial nations have an average urbanisation rate of 72 per cent. Less industrialized countries have 34 per cent. In the Asia-Pacific region the rate is 30 per cent, in Latin America 72 per cent, and in Africa 33 per cent. The urbanisation growth rate in a number of Asian countries has in fact slowed compared with earlier years. Nevertheless, not only industrialisation, but also the increasing degree of urbanisation has emerged as a growing burden on the environment in many Asian countries.

Changed Urbanisation Pattern in India

Environment burdens are just as much a problem in the old industrial nations as they are in India. But each group has a specific pattern of development. The urbanisation process in India has proven to be more pollution-intensive than that in the old industrial nations of Europe and North America. There are several reasons for that:

- industrialisation in India is restricted to a few locations which are often concentrated in and around capital cities. Although environmental damage continues to be minor at a national level, these locations have higher pollution levels than those ever reached in Industrial nations;
- furthermore, besides the strong regionalisation of industries, the industrialisation pattern of India shows a great diversity of environmental hazards. The trend to establish "last industries first", which is promoted by progressive industrialisation, leads to a country producing certain dangerous materials before they have been covered by state regulations;
- the time factor has to be seen as an important element in the emergence of these already highly regionalized environmental burdens. In India industrialisation and its concomitant urbanisation is taking place within a ban population grew tremendously.

Growing Environmental Damage

Water pollution in India is caused mainly by domestic sewage. For example, households are responsible for 75 per cent of the pollution of the rivers. The domestic sewage problem got more and more out of control with growing urban population. Pipe-based waste water systems are rare in this country. In India dealing with waste has an extremely low priority. The type of waste disposal depends mostly on what the cities can afford. The present level of air pollution is also very high.

Innovative Approaches to Solutions

Environmental protection and economic development are seen as contradictions. Economic development can only be achieved at the cost of higher levels of environmental pollution. And in reverse, if pollution is to be controlled and reduced this can only be done to the disadvantage of further development. In the meantime, however, numerous instances of successful urban environmental management are developing. They could help to change and subsequently break through the existing pattern of thinking. The following approaches can be viewed as important.

Combining regulations with incentives: The introduction of lead-free petrol and the mandatory equipping of new cars with catalytic converters is still by no means common in India. As numerous cars without catalytic converters are still able to use lead-free. Converters were then at first made compulsory for higher-powered cars, and later also for compact models.

Combining regulations with simple controls: Apart from general limitation of the number of cars in the city, its most important single measure to prevent traffic jams and the additional petrol consumption and pollutant emissions caused by them.

High economic growth in India has in fact led to a general reduction of poverty. But the distribution of income, particularly between urban and rural areas, has remained relatively constant. Urban environmental and traffic problems have increased heavily during the same period. These developments can be attributed to a certain pattern of official action (or "non-action"):

- governments have made efforts in supplying roads, but neglected the demand for mobility.
- governments are preoccupied with supplying water, and have neglected follow up problems, above all the questions of waste water disposal and treatment. In Indian cities, for example, this leads to the absurd situation that due to the mushroom like growth of the cities and the increased water pollution linked with it, water must be brought in over ever greater distances and at ever greater expense;

- governments take a one-sided look at noxious substances. Concentrations of harmful substances in water and in the air are in fact checked, and some measures are taken against individual pollutants of single sectors (e.g. lead emissions by the transport sector).

But an integrated policy which operates integrated environmental management with the aim of comprehensively relieving the burdens on the environment has not yet been developed anywhere. To consider such a concept, it is necessary to cut loose from the customary way of approaching problems. It makes sense not to separate the problem areas from each other according to sectors and pollutants, but rather on the basis of their ecological impact.

Orienting on demand hits the core of the concept of ecological modernisation, which is about reducing the intensity of resource use (note, at this stage this does not yet mean the absolute reduction of inputs). At the same time, sights are set on a lower use of land with the same size of population, or also lower energy consumption with the same degree of added value or the same per capita income.

Finally, the importance of governments for creating framework conditions must be emphasised once again. Because the actors come from different spheres, such conditions are essential.

From the time of the Greek polis, it was the ambition of the Greek city councillors to pass on a city that was more beautiful than the one they had taken over. There is a long way to go before such an attribute asserts itself in India (and elsewhere).

19

Towards Healthy Cities

More than a third of the urban population in developing world live in housing of such poor quality with such inadequate provision for water, sanitation, drainage, garbage collection and health care that their health is constantly under threat. But, properly planned, cities can be safe and healthy.

In the cities of India, it is common for one child in three to die before the age of five and for virtually all infants, children and adults who survive to have disease burdens many times higher than they should.

Diarrhoea, tuberculosis and respiratory infections (each among the largest causes of death) are generally much increased by over-crowding. Many accidental injuries happen when there are three or more persons living in each small room in shelters made of flammable materials and there is little chance of providing occupants (especially children) with protection from open fires or stoves.

But cities also include some of the India's safest and most healthy neighbourhoods. High densities allow much lower costs for supplying each household with piped, treated water supplies and most forms of health, educational and emergency services.

Sanitation and drainage may be costly in cities, as complex systems are needed to cope with high densities and large population concentrations but city households can generally afford to pay more—and are prepared to do so if they get a good service.

Cities may be considered ecologically unsustainable because of high consumption and waste levels but well planned and managed cities can combine high living standards with remarkably low levels of energy consumption, resource use and wastes. The concentration of people and production creates many more possibilities of collecting and recycling wastes and for walking, bicycling and a high quality public transport.

For many, city-life is one of excessive workloads and drudgery, yet cities remain centres of culture—including the visual and decorative arts, music, dance, theatre and literature. Most cities have a large reserve of young people on whose initiative and energy they could draw to improve condition—yet most such people find that their cities offer them little hope and little prospect of employment. If cities have such potential to provide healthy, stimulating and valued places to live and work for all age groups, why do so achieve this?

Supporting Change

Much of the explanation is the lack of 'good governance'. Good governance in any city means encouragement and support from all levels of government for a great range of investments of capital, expertise and time by individuals, households, communities, voluntary organisations and NGOs—as well as private enterprises. In most cities in India, the total value of investments made by people in their own homes and neighbourhoods exceeds many times the total value of capital investments made by city and municipal authorities. Yet governments and aid agencies usually ignore (or deem illegal) most such efforts.

Most households who want their own home cannot afford to purchase one—or at least one that is legal. They cannot obtain housing loans so the cost of the house purchase can be spread over a number of years—as they cannot meet the (usually) inappropriate conditions set by banks or housing finance institutions. If they turn to building their own home—as most do—they have to occupy or purchase the site illegally. They often have to build on dangerous sites—in floodplains or on slopes with frequent landslides or mudslides—as the cost of safer sites is too high.

Even if they can qualify, for a housing loan; most such loans are for finished houses, not for incremental construction. And even when they have developed their own home and neighbourhood into a viable residential area, governments usually refuse to provide these with roads, water supplies, drains and other essential infrastructure, because they are 'illegal'.

What would cities look like today if governments had supported these individual and community efforts by ensuring that land, building materials, credit and technical advice were as cheap and readily available as possible? Or if government-community partnerships had been formed to, at least, improve water supply, sanitation, drainage and health care.

These work within what is often called the 'social economy'—the great variety of initiatives and actions that are organized and controlled locally and that are not profit-oriented. The social economy includes the work of citizen groups, resident's associations, street or barrio clubs, youth clubs, and parent associations that support local schools. It includes many voluntary groups that provide services for the elderly, the physically disabled or other individuals in need of social. It often includes many initiatives that make cities safer and more fun-helping provide supervised play space, sport and recreational opportunities for children and youth. It may provide formal or informal supervision or maintenance of parks, squares, and other public spaces.

The social economy not only 'gets things done' but also creates a dense fabric of relationships that allows citizens to work together in identifying and acting on local problems.

Its value to a 'healthy city' is enormous, even if it is often forgotten by governments and international agencies.

The capacity of city authorities to govern is not the same as the capacity to invest, since these authorities can do much to encourage and support the social economy. City authorities can often greatly increase the supply and reduce the cost of land for housing by changing inappropriate regulations, streamlining planning and land use control, procedure and making better use of publicly owned land.

City authorities should also have the main rŏle in enforcing legislation on, air and water pollution and occupational health and safety. This does not require large investments by public authorities, but it can do much to improve health and the quality of life in a city. Good governance also means managing competing claims and finding common ground between enterprises, trade unions and residents about what should be done to make the city more healthy.

Achieving a healthy city needs a representative political system through which the priorities of citizens and businesses can influence policies and actions. Democratic structures remain among the best checks on the misallocation of resources by city and municipal governments. Actively involving a wide range of local groups in developing 'city governance' helps ensure that the different priorities of a wide range of groups are addressed.

The key issue is not so much identifying what should be done to achieve more healthy cities. This is well known. It is identifying how it should be done, especially how governments and international agencies can support a vast range of activities by individuals, households and communities that help build and maintain healthy cities—which to date they have ignored or even (for many governments) repressed.

20

Sustainable Cities

Today almost one half of the world's population lives in cities. The world's cities are growing by one million people each week. Cities today play a significant role in development. They continue to attract migrants from rural areas because they enable people to advance socially and economically. Cities offer significant economies of scale in the provision of jobs, housing and services, and are important centres of productivity and social development.

However, the stress of this rapid urban population growth is often overwhelming. The long list of afflictions includes urban poverty rates of up to 60 per cent. Despite growing investments, more than one-third of the urban population live in substandard housing. Forty per cent of urban dwellers do not have access to safe drinking water or adequate sanitation. Primarily due to a rapid growth and a deteriorating urban environment, at least 600 million people in human settlements (cities, towns and villages) already live in health- and life-threatening situations, and almost 50 per cent of these are children.

The high rate of urban population growth in most regions has led to common problems: congestion, lack of funds to provide basic services, a shortage of adequate housing and declining infrastructure, to name a few.

While these problems are occurring in urban areas, cities still have an important role to play in protecting the global environment in the face of rapid urban population growth. Agricultural and livestock production in rural areas are pushing farther and farther into ecologically fragile regions and cannot support growing population. The finite land and water resources make it imperative that human settlements be carefully planned. Indeed, sustainable urbanisation will ease the pressures caused by encroachment on fragile natural habitats.

India's cities offer a bewildering sight to any visitor: the congestion caused by rapid population growth and a continuing rural-urban drift often leads to conditions which defy all rules of orders, hygiene and environmental safety. Inadequate leadership, corruption and mismanagement have a harmful effect on the physical, environmental, social and ethical structures of cities in India.

Millions of people live in inadequate conditions—without piped water, electricity, security of land tenure, access to roads or health facilities. The means available for production and financing of housing and urban infrastructure are too limited to meet basic needs.

Reducing Poverty and Creating Jobs

Urban poverty is rising at an alarming pace, especially among women. The informal economic sector—which makes a substantial contribution to the delivery of services, production of goods, building of infrastructure and housing construction—often provides the only opportunity for the urban poor to make a living.

Local informal housing construction, for example, generates up to 20 per cent more jobs than high-cost construction. Street hawking, waste recycling and food production are primary sources of income among the urban poor and are illustrative of the creativity of survival strategies.

However, the informal sector itself is often highly exploitative and fails to raise people's economic development beyond mere subsistence. Larger economic strategies and more participatory urban planning approaches that take stock of local skills, technologies and materials are required to generate new and better-paying job opportunities in cities and towns.

Incorporating Environmental Concerns

In 1992 the Rio Conference on Environment and Development designed the Agenda 21 Programme of Action to help save a planet endangered by environmental neglect and plagued by poverty and underdevelopment. Most of the goals agreed to in Rio can become reality only through local action in cities where environmental threats are increasing. Again, it is the urban poor who are particularly endangered by environmental degradation and pollution. The world's Agenda 21 will fail if the city's environmental agenda (population, inadequate sanitation, water supply and waste management) is not addressed. This is being recognized by local authorities all over the world.

Sustainable development in the twenty first century will to a large degree, depend upon how cities, towns and villages everywhere interact with the environment and utilize natural resources.

Increasing Awareness of Gender Issues

Women and men use and experience cities differently, according to their roles, responsibilities and access to resources. For example, when basic services are lacking in a settlement, more often than not it is women who take on responsibilities such as water collection and refuse disposal. Women often have unequal access to resources such as property, credit, training and technology. All of these factors must be addressed urgently, as they make it harder for women to improve their living standards and those of their children.

Disaster Mitigation Relief and Reconstruction

As cities become large and more densely populated, they become increasingly vulnerable to natural and man-made disasters such as earthquake, floods, industrial hazards, epidemics, civil strife and wars. Poor people are forced to live in the most exposed, dangerous and cramped conditions; in flood-prone areas, on steep hillsides or near polluted streams and waste dumps. As a result, they are most likely to lose their homes or their lives when disasters occur. Better planning, access to affordable urban land, and improved construction methods can reduce the extent of catastrophes.

These successful and sustainable approaches to poverty eradication; managing the urban environment; providing access to land, shelter and finance; empowering women and men; and many other issues will have to be documented and disseminated widely.

21

Consuming the Future

Now that we are to reach six billion of us, it is a good point to check again on what sort of lifestyles we pursue and what is the environmental impact of those lifestyles. It is curious that we have spent several decades being concerned about the growing numbers of humankind while not giving at least an equal amount of attention to the levels of living we aspire to, and how many natural resources we chew up thereby and how much pollution and waste we cause.

Everybody is a consumer of sorts. True, every fifth person scarcely qualifies for that designation, consuming goods worth less than $1 per day. Conversely, every seventh person qualifies for a designation of super-consumer, with a cash income at least fifty times greater. These latter are the people who, through their carbon dioxide emissions, are disrupting everybody's climate dozens of times more than the average citizen of One Earth. Fair play, anyone?

Much as the have-nots seek to match the have's, it is plain their efforts will not work out for a long time to come, at best. If every Chinese person were to consume just one additional chicken per year and if the said chicken were to be raised primarily on grain, this would account for as much grain per year as all the grain exports of the number two exporter, Canada. If the Chinese were to raise their per capita consumption of beef, now only 4 kgs per year, to that of Americans, 45 kg, and if the additional beef were produced largely in feedlots after the manner of the United

States, it would account for as much extra grain as the entire US grain harvest, less than one-third of which is exported. Because of its recent climbing up the food chain toward a meat-based diet, China has become one of the world's leading importers of grain. The global grain market today is around 200 million tons per year, and shows scant scope for significant increase.

As a further measure of its ambitions, the Chinese government has designated the auto industry as one of five industry "pillars". Today China has fewer cars than Los Angeles. If per capita car ownership, together with oil consumption, were to match that of the United States, China would need 80 million barrels of oil per day—by contrast with the world's 1996 oil output of 64 million barrels of oil per day. The surge in carbon dioxide emissions would be unprecedented.

All this notwithstanding, there are already some 250 million newly affluent people in China. They are people with a household income equivalent to perhaps US$20,000, and enough discretionary income to enjoy the per quisites of the good life as perceived by these nouveaux riches. Top of the shopping lists are meat and more meat, followed by cars whether big or small. These are the badges of success: they show you have arrived.

The new consumers in China are matched by at least 200 million in India, and tens of millions in South Korea, Taiwan, Malaysia and Thailand (the recent economic setbacks have not permanently punctured the economic bubbles). Then there are 200 million more in Brazil, Argentina, Venezuela and Mexico, and more again in Hungary and other countries of Eastern Europe, also Turkey. Put them all together and they total about as many as the 800 million long established consumers in the ultra rich countries (the OECD grouping). When the current economic hiccups in Asia are left behind, the ranks of the new consumers can be expected to rise rapidly.

But they cannot hope to become super consumers. Where would all the extra gain come from? How could the global climate tolerate the huge additional pulse of carbon dioxide? There are all kinds of other environmental reasons to suppose that environmental constraints will become all the more constraining. True, technology could help moderate the environmental impact. We could enjoy twice as much material prosperity while using only half as much natural resources and causing half as much pollution and waste. But the new consumers will want to pursue the American dream to the hilt, and it is hard to see that the best technologies could enable huge numbers of affluent aspirants, perhaps two billion people by 2010, enjoying even half the material prosperity of Americans with average household incomes of $40,000.

But is it true "prosperity"—mental and emotional as well as material? Or is the American dream becoming a nightmare with its harried lifestyles and declining leisure time, where the shopping mall is the ultimate mecca, and the good life is a case of piling up goodies?

In any case, we cannot expect the new consumers to forego their "rightful share" of affluence unless the long-time affluent agree to cut back on their environmental ruinous lifestyles. It is these communities that must offer a strong example, and soonest. Where is the political leader who will espouse the new vision, however much it may be perceived as the ultimate vote loser?

22

The Future of Work

The advent of an 'intangible' economy does not mean the end of work. But it does mean the end of familiar routines and rhythms, of job security, of rigid hierarchies and career planning.

People are worried about the far-reaching transformation of the economy. Are we heading for "the end of work". Yes, we have reached the end of the road. We are no longer creating jobs in industry and automation is sure to reduce their number in the services sector. The quantity of work is thus inexorably bound to decrease.

This thesis may be popular, but it is also mistaken and harmful. History shows that technological innovation has always created jobs on a large scale. In no way is the current trend leading to "the end of work". Just the opposite; the new economy contains huge pools of new jobs which can more than make up for the inevitable loss of traditional jobs.

Dematerialisation—the shift away from material products—is revolutionizing all aspects of work—its nature, its organisation and its relationship with other activities. Its function is no longer just the manufacture of physical objects but the handling of data, images and symbols. The content of jobs is becoming more abstract. Skilled workers need to know a lot more about mathematics than their fathers or grandfathers did. Even milking cows and manufacturing require more and more calculation, evaluation and control.

Financial Markets that Never Sleep

The organisation as well as the product of work is also becoming increasingly intangible. The unity of time, space and action which characterized work in the industrial economy has disintegrated. Work is no longer a regular eight-hours-a-day, five-days-a-week routine. New rhythms have appeard—the hectic pace of financial markets which never sleep, the ups-and-downs of life in show business and the uncertainties of "just-in-time" production where components are delivered a few moments before the final product is assembled.

The new jobs are quitting familiar workplaces such as factories, offices and warehouses. Telework is increasing. Europe's teleworkers may number 10 million by the year 2000, up from one million in 1994.

This upheaval of worktime and workspace is going hand in hand with a functional explosion. The range of skills and types of work is expanding all the time. In the United States, the number of job categories has risen from eighty in the 1940s to nearly 800 today. At the same time, trades are dying out faster and faster, especially in information technology where many jobs have a short life of only a few years. Jobs are becoming simultaneously more evanescent and more pervasive, more dissociated and more integrated. On the one hand, fragmentation in time and space seems to be more extensive than it was in the industrial economy. On the other, information technology is strengthening the links between different stages of work and creating an overall fluidity.

Disparities in Productivity

The new forms of work are non-linear. When handling information, knowledge and feelings, there is no direct relationship between the amount of efforts and the final result. This makes of very wide disparities in productivity. In industry, the ratio of the performance of an average worker to that of a good one is no more than one to five. But in immaterial work, an excellent programmer can be a hundred times more productive than an average one.

Non-linear work means non-linear organisation. The notion of a rigid, formal hierarchy based on unchanging criteria no longer makes much sense. All that matters now is technical, scientific or artistic skill and the ability to establish a solid relationship with the customer. Functional hierarchy is replaced by "brainpower"—authority gravitates to those who create and control the new stock of intangible assets: data, brand image, technological know-how and human capital.

The new techniques for managing human resources are individualizing the assessment of performance. Two people doing the same job may have different salaries and different status. Automatic across-the-board pay rises are being dropped and replaced by bonuses linked to results. There are no sinecures in the new business enterprise, either for rank-and-file employees, supervisors or technicians—the supposed beneficiaries of the new knowledge economy.

Business leaders are no longer a protected species. The head of a big American firm is ten times more likely to be sacked for poor performance now than was the case twenty years ago. The notions of loyalty and of indissoluble links between a firm and its employees are losing their meaning.

The changing nature of work has led to a big increase in so-called non-typical jobs, including part-time, temporary and flexi-time work and short-term contracts. Almost all the jobs created in Europe between 1992 and 1996 were part-time. This trend worries many observers who see it as hidden under-employment or disguised unemployment. But they are overly pessimistic. The growth of non-typical jobs is the result of the convergence of several persistent developments.

Where the New Jobs Are?

The shrinking number of jobs in traditional sectors of the economy seems to be a general and irreversible trend. In rich countries as a whole, the share of industrial jobs fell from 28 per cent in 1970 to 18 per cent in 1994. Meanwhile, the share of the services sector grew steadily. Four major new sources of jobs can be identified:

Handling Information and Knowledge: Computer services, research and development, teaching and training account for 40 per cent of knowledge workers. These high-intensity knowledge activities comprised 43 per cent of all new jobs created in the United States between 1990 and 1995, but only 28 per cent of total jobs.

Information Technology: Here there is a shortage of personnel. Professional groups are sounding the alarm and calling on governments to help. In the European Union countries, the imbalance between supply and demand is such that half a million jobs are waiting to be filled.

The Health Sector: The growth of high-intensity knowledge services in this field is related to increased life expectancy and the ageing of the population, and the demand for physical and psychological well-being is also steadily increasing. The growth of expenditure on health is persistent and widespread. For the OECD countries as a whole, this spending grew from 3.9 per cent of GDP in 1960 to 7.2 per cent in 1980 and 8.4 per cent in 1992.

The Leisure Economy: This has triggered the expansion of cultural, sporting and leisure services. It ranges from amusement parks and rock concerts to cultural events such as opera and major art exhibitions. The products of the culture industries have become mass consumer items. Never before have people read so much, listened to so much classical music or visited so many museums. Information technology is also going to add to this vast range of consumer choice. In sourthern California and New York, the entertainment and multimedia professions are among the main sources of new jobs.

In the labour market, the increase in non-typical jobs is one of the ways in which employers are responding to the pressures of competition and adapting to a global economy which functions seven days a week, twenty-four hours a day. To cope with the new situation, firms are having to figure out how they can use their workers more efficiently and flexibly.

The growth of non-traditional jobs is also due to changing demand. Consumers want to be able to buy a very wide range of goods and services at the drop of a hat, or amuse themselves any time, anywhere. To meet this demand, shops and places of entertainment have to be open late at night and on Sundays. Technology encourages this trend: the virtual economy of the Internet never sleeps.

The widening range of types of work also reflects long-term demographic trends, especially the greater number of women workers and longer life expectancy. Some see non-typical jobs as a necessary evil, while others, especially women with children, welcome the change.

The divide between traditional kinds of work and the new jobs is no longer watertight. People are increasingly switching back and forth between the two categories. In the course of a lifetime, a person may change from full-time to part-time work, from an office job to home office and from the security of a big firm to the adventure of entrepreneurship.

Changes in the nature of work are also breaking down the rigid frontiers, which marked off the world of work. The traditionally distinct fields of work, education and leisure are now interwoven and coexist flexibly in a kind of triple helix of social life.

The emerging intangible and relational economy has a huge potential for growth because it is not bound by the constraints of material scarcity. However, the transition to the now economy is an open-ended process. The state has a key part to play in bringing it about. Governments can slow down the rate of change by making it more painful and more costly.

Obstacles to Change

Pessimistic scenarios are still plausible, such as that of an economy which generates few new jobs and is polarized between a small elite and the rest of the population who are marginalized and lie in precarious conditions. There is a big risk that this scenario will come to pass because current laws and regulations, as well as widespread pessimistic ideas about work, are powerful obstacles to change. Optimistic scenarios require a wholesale reform of institutional structures and profound changes in behavior and attitudes. Such far-reaching changes often run into strong opposition from the social and political establishment and come up against the weight of psychological and social tradition. But the gamble of a new approach to work must be made if the transformation to the intangible economy is to succeed.

23

Energy and Sustainability

Mankind's history is marked by a growing use of energy which until the end of the Industrial Revolution came largely from renewable sources. It was coal that fed the furnaces and boilers of the Industrial Revolution from the end of the seventeenth century to the nineteenth century, and drove railway transport and steamships. As well as being a useful source of mechanical energy, it was also used in the manufacture of coal gas for street lighting and in the chemical industry. In fact, coal was the principal form of energy until 1900.

Discoveries at the beginning of the nineteenth century allowed the use of electricity and revealed the relations and the interconvertibility of different forms of energy. The principles of conservation and of energy quality did not become operative until much later. Meanwhile, in 1882, the first system for producing and distributing electricity in a large city was installed. This was the beginning of the second phase in industrialisation through electrification.

Following the first successful oil drillings in 1859, Standard Oil, the first of the modern large scale oil companies, attempted the first vertical structure for overall control of the oil process. It involved extraction from the subsoil, storage, refining and final distribution. Later, the growth of derivatives, the lower extraction costs compared to coal and the greater ease and economy of transport made oil modern society's basic energy source.

The internal combustion engine led to motorisation on a massive scale by land, sea and air and guaranteed a constantly growing market for petrol. The forties marked the start of the new petrochemical industry, which gave rise to an enormous number of new products; synthetic rubber, plastic, medicines, cosmetics, varnishes, artificial fibres, detergents, weedkillers, fertilizers, butane, propane, etc., opening the way to the mass-production of consumer goods and introducing new, non-biodegradable substances into the environment.

After World War II, ambitious programmes to produce electricity from nuclear energy were begun, in the search for a return on the enormous amounts of money invested. The economic expansion in the West during the fifties and sixties was directly related to enormous petrol consumption at a time when energy was considered plentiful and cheap. Energy consumption during these decades grew more than exponentially. The fastest developing industrial sectors were precisely the ones that consumed most energy—petrochemical industries, metallurgy, car manufacturing, domestic appliances, electricity generating, etc.—and a trend developed towards goods and services with higher energy intensity. Since 1950, increased energy production has been systematically favoured over more rational use. So much so that the increase in energy consumption has been taken as a reliable indicator of progress.

The Aftermath of the Oil Boom

The oil crises of 1973 and 1980 showed up the fragility of an energy system that was over-dependent on oil. The War in the Gulf was reminder of what was at stake for the Western economies; free access to cheap oil in the Middle East. It was therefore fear of the hardship caused by the first crisis that brought about a change in attitudes in Western countries; efforts were directed at breaking free from this dependence, diversifying supply sources, perfecting replacement energies and promoting energy-saving programmes.

The eighties marked a change in people's awareness about environmental problems. The damage was making itself felt in more and more places and eventually the global threat to our planet as a result of our energy system became clear; the composition of the atmosphere was changing and could lead to possible changes in the climate.

According to recent figures, 82% of all the energy consumed in the world is produced by burning fossil fuels, 7.5% from burning biomass, 5.5% from the use of hydraulic energy and 5% from nuclear energy. Most of our energy in other words, in non-renewable; it runs out as we use it, as the population increases; and it comes from fossil fuels, which on burning increase the amount of CO_2 in the atmosphere. If we add to this the accumulation of nuclear waste, the problems of access to oil deposits, constant spillages during transport and all the different imbalances involved in the world energy system, the outlook is far from sustainable.

The inequalities speak for themselves; globally, less than a quarter of the world's richest population consumes almost three quarters of the energy commercialized in the world. For example, the average annual consumption per capita in the United States is 26 times higher than in India.

The Choice of Change

Opening the way to societies that make sustainable use of energy necessarily involves increasing and improving energy efficiency, both in supply technologies and in end-use technologies, at the same time using renewable energy sources instead of fossil fuels.

Choosing the right system for the transformation of primary energy sources into energy services such as lighting, cooling, cooking, mechanical force, transport, etc. and choosing the most suitable appliances and technologies in each case is fundamental.

The truth is that a good standard of living is possible without wasting anything like as much energy. A series of relatively straightforward measures today allow a far higher level of comfort than in 1950, using one-third as much energy for heating water for washing in the home.

Petrol consumption by vehicles has dropped by 40% in forty years, from 8 litres/100 kilometres to 5.3 litres in some models, and the work of improving their energy efficiency continues. In industry, the energy consumption necessary for manufacturing large intermediary products (steel, cement, paper or fertilizer) is decreasing steadily at a rate which varies between 0.5 per cent and 0.2 per cent per year according to the product.

Today's incandescent bulbs consume one twentieth as much electricity as bulbs in the twenties. The compact fluorescent bulbs now available can cut this down again to one-fifth. Efficiency in lighting has increased one-hundredfold. The use of new materials and a more rational use of traditional materials allows a reduction in the amount of energy and raw materials consumed. Building a house, for example, requires 20 per cent less energy than in 1950; building a vehicle, 40 per cent less. On a global level, reducing our society's energy-intensiveness is the first step towards energy sustainability.

24

Development

The Third Way

While great claims are being made for the increasingly more efficient and effective technologies perfected to serve development of the people in this scientific age, huge problems are threatening the globe. The problems are mass poverty and hunger, underdevelopment, waste, unemployment, resource scarcity, environmental destruction and armed conflict.

In finding answers to the prevailing problems we must be clear about the meaning and purpose of development. The first glaring mistake made is that development is interpreted as development of the economy and not as the total development of society. When economic development is made the supreme goal, most of the other vital aspects get ignored, namely, development of the political system, community, social cohesion, the ecology, culture and values. Development and stability go hand in hand while poverty and chaos constitute the antithetical twin.

Appropriate Development

The key elements in the conception of appropriate development consist of: first, aiming at sufficiently comfortable material living standards and not affluent standards of the rich as in prosperous nations. Second, development must not be confused with GNP growth. Mere increase of economic activity must not be pursued exclusively at the cost of articles that are urgently needed by the poor majority to maintain them at a reasonable level of material living. Third, in the villages, we must produce articles as are needed by the villagers. Fourth grassroots and participatory development is essential so that the local people identify and solve their local problems. Fifth, instead of capital and energy intensive high technology, use labour-intensive technology. Instead of heavy industrialisation, promote medium scale industries and technologies. And, sixth, instead of preoccupation with a high GNP growth rate, focus on the development of communities and of rural bodies and take care to conserve the local ecosystems. The main purpose should be to meet their primary needs of ordinary people and promotion of their productive resources such as land.

In developing countries like India where billions of poor people remain condemned by conventional economic development strategies and theories, it is vital to introduce appropriate development measures to remove deprivation and ensure the necessity of modest living standards.

The world has witnessed the operation of the two systems namely, the capitalist and the socialist one that have obtained in different countries. Although both the systems have underlined the welfare of all as the basic goal; both have left a legacy of waste, hunger and gross human inequality. Some 1000 million people do not get enough to eat including some 20 million in the USA.

Third World Way of Development

The iniquitous situation in the present day world has sparked off fierce controversies among the conventional economists and the new radical economists who champion a third way, as the alternative way to serve the primary goal of all humanity to have sufficient means to lead a comfortable peaceful life.

In order to achieve prosperity, conventional economists have emphasized the production of bigger cake on the assumption that everyone will get a slice of it. They also argue that a "tide will lift all the boats". Both these assumptions have proved false in that the poor have neither the slice of cake nor has their boat been lifted. Third way system lays stress on highly localized, less cash-reliant and simply structured set-up. It should not be dependent on transport of goods, but concentrate on more local production to meet local needs with a role for barter and free exchange. He urges a radical re-think of conventional economics.

Is this Stepping Backwards?

The most common criticism levelled against the Third Way is that it will arrest the progress made hitherto and that it might mean a return to a 'primitive' way of life. There should be no fear on this score because the Third Way aims at the reduction in the use of resources and therefore, of excessive production and consumption. It does not in any sense mean stepping backwards to a lower level of the quality of life. Nor is the alternative way intended to destroy capitalism or socialism. The conception of the alternate way is to promote economic growth compatible with capitalism and socialism. The ground idea is to promote selflessness, mutual concern and social responsibility. This will replace selfish, competitive and avaricious attitudes as have developed in the conventional economic order of today.

NGO's Resolution at the Rio Conference

That there is increasing awareness of the threat posed by the growing power of multinational corporations was articulated by the international NGO forum in its Declaration resolved on 12 June 1992 at the UN Conference of Environment and Development in Rio de Janeiro. The Declaration states that "the Bretton Woods institutions have served the major instruments by which the destruction policies have been imposed on the world" and calls upon "the world's people to protect their economic, social, cultural and environmental interests against the growing power of transnational capital". The declaration further avers "we recognize the central place of spiritual values and spiritual development... and values of simplicity, love, peace, and reverence for life.

After the failure of the socialist system over four decades to achieve prosperity for all, the Indian Government switched over to the global market economy and is steaming ahead with added liberalisation measures to attract foreign investment and the multinational corporations (MNCs). Some adverse effects of this are already visible: for example, majority financial equity granted to MNCs and the emergence of foreign subsidiaries with cent per cent financial equity; the introduction of pizza and Kentucky Fried Chicken which has been detested by the people in Karnataka. The farmers have also revolted because their rights to produce and sell seeds have been wrested by foreign MNCs who have acquired patent rights over certain Indian crop seeds. In this scenario, the third Way has much to commend itself to the Government. The third Way has the air of the Gandhian model of economy and production which emphasizes production by people for their own needs and preference for small and medium sized industry. The same paradigm was championed by the renowned economist Schumachar when he said "Small is beautiful". India should take good care against the present-day headlong drive for the entry of foreign capital and foreign heavy industries.

25

Employment and Promoting Ecology

How a Service Culture Could Put People Back to Work

We are facing two big and urgent social problems: employment and ecology. Both the unemployment of millions of people and the progressive destruction of the ecosphere are alarming. But they are linked with each other. The 'greening' of industrial products, processes and services could provide many more jobs.

Unemployment has many causes, including:

- Sluggish markets;
- Stagnating or declining purchasing power;
- Growing uncertainty about the future at all levels;
- Lack of will and/or ability to innovate.

But joblessness is by far due mostly to the high efficiency of industrial machinery, which produces ever more, ever faster, with ever fewer workers.

Waste of Resources

The extremely high productive use of human labour and the extremely low productive use of resources are manifested by gigantic mountains of waste. Already today, the junked cars on scrap heaps alone would form a line that would reach to the moon. The scene is the same with discarded electrical and electronic appliances. Every year, millions of tons of ovens, washing machines, refrigerators, dishwashers, TV sets, entertainment electronics equipment and small appliances are being wasted.

If we throw away all these things after a relatively short time we are not only being wasteful and irresponsible with resources, but equally so with people's work. For with the products and materials we discard, we also dispose of the human labour they contain. It is imperative that we radically reduce the enormous turnovers of material and energy. In other words, the productivity of raw materials and energy must be markedly increased. Specifically, that means we must draw as many services as possible from one kilogram of material or 1 kWh of energy. Reducing the enormous flows of materials into the industrial system, as well as developing cycles of materials and responsibility (the manufacturer taken back and repairing and/or remanufacturing used products and materials) and the main pillars of a sustainable development that can cope with the future.

The industrialized nations must cut their consumption of raw materials by a factor of about 10 by 2050; if they are to be able to handle the challenges of the future. To achieve that reduction, innovation efforts must be directed at increasing resource productivity and/or ecological efficiency. In particular, strategies to extend the useful life of goods and intensify their use could result in reducing both the speed and volume of the flows of resources to industry.

Increasing Resource Productivity

In dealing with nature, we and industry are facing radical change. This is the transition from environmental protection (preservation of nature and health) to greater resource productivity (which at the same time means greater competitiveness). As a rule, environmental protection costs money, while higher resource productivity usually cuts manufacturing costs and/or increases a company's profitability. If the company can sell the same utility or benefits while using fewer resources, it saves twofold: in buying raw materials and on waste disposal. Thereby the rule is that goods and components cycles are more profitable than resources cycles, and that the company which is first in the market gains an additional competitive advantage in terms of a lead in knowledge and image. If a service, or benefits in the form of services, can be sold instead of products, the decoupling of company success and materials flows is even greater.

Impacts on Employment

The two social problem areas of work and ecology have to date been perceived and treated separately in politics, in industry and in our own minds. And, I believe, with little result. The link between the two must be established.

The strategies to boost resource productivity would have considerable impacts on the change in industrial structures, on handling existing product inventories, and on employment. In particular, the strategies would lead to a switch of focal point from a raw materials-intensive and use-value-related service economy. This is where another view of profitability comes in. Business management would no longer focus on value added, but on maintenance of value over longer periods based on the intrinsic value of a product. Expressed as a question, the value factor, which would move to the centre of business thinking and dealing, means: how can the utilisation value be improved and sold? How can products be made with as few raw materials and as little energy as possible and create a high benefit as pollutant-free as possible for as long as possible during their entire life-cycle?

With regard to employment, the production of long-life goods would appear at first sight to lead to a reduction in the need for work. In fact, however, the strategies to increase resources productivity have positive net employment impacts.

The reason is that saving resources is based in principle on substituting energy by work, rather than the reverse as has been customary to date.

If the useful life of products is extended, that will not only preserve most of the materials and energy they contain as well as the work invested in them. The products will also require a considerable amount of mostly skilled work input. Reconditioning products is as a rule more labour-intensive than manufacturing them. So large-scale reconditioning and repair work increase the number of skilled jobs and at the same time reduces the inflows of materials and energy.

Comparing a car with a life-cycle of 20 years with two others that each have useful lives of 10 years gives a good example. The first car causes an increase in employment per life-year of about 50 per cent in terms of total work input in manufacture, service, repairs and reconditioning while at the same time reducing the energy consumption by half.

Regionalisation of Industry

Extending product service life would also mean replacing energy and/or capital by skilled work, helping to save money to boot. But not only rising costs of disposal, materials and energy would reduce consumption. Increasing transport costs would also mean that carrying all kinds of freight halfway around the world would make less and less business sense. That would result in ever more products and materials being circulated, reconditioned, and recycled or reduced on a regional basis. In turn, that would create regional jobs, and be more profitable as well as more promotive of technology—not only from ecological aspects.

In addition, a way of doing business which encompassed material and responsibility cycles would no longer differentiate between manufacturing and reconditioning, or between marketing and remarketing. The structure of such an economy would be predominantly decentralised and regionalised so that it could adapt itself to the new cycles. It also would benefit from the greater efficiency of the new working practices.

True, jobs would be lost in the sectors of central production, and raw materials extraction and processing. But at the same time, more and higher-skilled jobs would emerge. These would not only be better qualified jobs, but also decentralized because reconditioning, repairs and maintenance must be done near the customer. And that, in turn, would also reduce goods traffic.

In addition, skilled workers would be needed because in many cases of small production runs it makes sense and is also more economical to hire such people. They can work faster and more flexibly and mostly cheaper than fully automated production lines.

There also would be a growing need for maintenance, repairs and reconditioning. More and more people would be wanted for reconditioning, that is, the remanufacturing of old products. As reconditioning involves far more craft work than highly rationalised new production, there would be a positive impact on the labour market if there were more of the former and correspondingly less of the latter.

From Production to Services

Switching to long-life products and changing from selling products to selling use-values would strengthen the current trend of jobs shifting from industrial production to the service sector. For example, if the service of individual transport were to be sold instead of the product car, the company with the competitive advantage would be the one that had a service centre in every town and village, with appropriately staffed workshops and sales or rental facilities.

Enduring change towards a knowledge-intensive and use-value-related service economy would not only mean that more people would be needed to fill jobs. It would offer more opportunities for part-time work, as well as possibilities of employment for older people and the handicapped. People who earlier could not keep up with the pace of working life would be more inclined to return to it. Another impact would be that many companies would reduce their dependence on the world market. They would no longer switch certain tasks abroad, but assign them to their part-time employees, helping them to meet their commitments as self-employed entrepreneurs.

The latter would be accommodated by an ecology-driven fiscal reform which would make massive cuts or changes in subsidies and raise the cost of energy and raw materials consumption. This move would be accompanied by a reduction in income tax and non-wage costs such as social security contributions. The market would thus be more efficient, energy- and material-intensive new production more expensive, labour intensive repair work and reconditioning cheaper, and jobs would remain in the home country or region.

A number of more recent studies show clearly that an ecological tax reform would help to create jobs, and thereby could make a decisive contribution to reducing unemployment.

26

South Asia Quarrels Over Water

Bangladesh, Bhutan, India and Nepal have at least three features in common; they are all situated on the southern slopes of the Himalayas, they have high levels of poverty and they are all rich in water resources. To these add a fourth: they frequently quarrel about water.

For decades, policy makers in the four countries have grappled with the problem of how to harness the rivers that flow from the mountains and serve the region's over-one billion people. Mostly, they have gone in for large dams and grand irrigation schemes. But though these promise a lot, their usefulness is being increasingly questioned by many independent engineers, scientists and social activists. And they are thought to contribute to water conflicts between countries.

Large hydro-power projects represent a development model that is biased toward industry and urban areas. And irrigation scheme favour rich farmers with large farmlands, most useful for water- and fertilizer—intensive cash crops that poor peasants cannot afford to grow. Because of their size, they also cause huge displacements-again, of poor farmers. Social activities say that while such projects may have increased overall food production, they have also increased the rich-poor gap in a region that is home to the world's largest number of absolute poor—people living on incomes of less than one dollar a day.

In tandem, many South Asian hydro-power engineers and economist say 'national' and top-down engineering solutions to water management ought to be replaced by a new 'basin-wide' or regional approach through environmental and socially benign projects.

One example of such 'national solutions' is the Indian Farrakka Barrage on the Ganges. Built near the Bangladesh border in order to divert water to the Calcutta port, it has led to problems with falling water tables and salinity downstream in Bangladesh. And although India and Bangládesh resolved the long-simmering dispute in 1996, Dhaka is now proposing another expensive Ganges Barrage to solve the problems created by the first barrage.

Of the four countries, Nepal has the highest per capita potential for hydropower generation—its narrow valleys and steep terrain mean water flows faster, which is good for power generation. But although Nepal could potentially turn its rivers into 'hydrodollars', Himalayan dams are expensive to build and the country's experience in dealing with India on joint river projects has been patchy. At the same time exporting power to India, where power-cuts that can last up to 12 hours at a time in some cities, means foreign exchange for Nepal.

Because many Nepalese perceive past border irrigation projects as unfairly benefiting India, joint Indo-Nepal river projects are a political issue. A recent treaty to harness the Mahakali River on Nepal's western border with India is a case in point.

Although India and Nepal agree on equally sharing the 7,500 megawatts of power from the project, negotiations have been stalled over Nepal's demand that it be compensated for downstream irrigation benefits to India. Nepal is not allowed to harness water in many of its river for its own use without India's concurrence these rivers flow into India, and Nepal is bound by past bilateral treaties.

As one official at Nepal's Water and Energy Commission put it: "What's in it for Nepal? Why should we write off benefits from the Mahakali for future generations of Nepalis? Thing look even bleaker for larger joint projects like a planned 10,800 mw dam on the Karnali River—the $10 billion Chisapani dam project is expected to displace some 60,000 Nepali farmers. In addition, India has already used up all the natural flow of the Karnali for its own irrigation downstream.

Things are different when it comes to Bhutan: India and the tiny Himalayan kingdom sorted out their river projects without any major difficulty although experts say it is too early to assess how much the Bhutanese will really benefit. According to an estimate by an Indian water consultant, Bhutan can generate up to 20,000 mw of power from its rivers.

Generating funds for large projects does not seem to be a problem. Traditionally, the World Bank has been one of the major financiers for dams worldwide having undertaken 400 projects involving dams since 1970. In 1988, India alone had about 473 dam under construction, with the Bank involved in 45 of them.

When the World Bank recently pulled out of some controversial mega projects, such as the Arun 111 dam in Nepal, multinational companies stepped in. In Nepal, three medium-sized hydropower projects worth more than 300 million dollars are being built with foreign direct investment.

FDIs are backed by organisations like the Japanese-managed Global Infrastructure Fund (GIF) which is interested in putting money into such projects as the 270 metre-high Kosi High Dam in Nepal, the Ganges Barrage in Bangladesh and the Mammoth 20,000 mw Dhihang dam in India. Funds also come from the Manila-based Asian Development Bank.

But many resource economists maintain that the chances of making colossal mistakes is higher with large projects. It is all a question of risk-management. Can our countries afford to take the risk of spending billion on doubtful projects that can go dangerously wrong?

27

Using Economics to Advantage

In the eyes of the public, the economic sectors—for instance energy, transport and agriculture—are often seen as pursuing interests that conflict with environment and health. They are the originators of pollution and often devise economic arguments to oppose changes in their practice that could improve environment and health. This behaviour has led the public, as well as environment and health professionals, to view economic analysis negatively. However, these economic arguments are often inadequate and unconvincing from the point of view of many economists.

In fact, the economic rationale is bound to reflect as closely as possible the preferences of the population and thus to take much greater account of environment and health. If used by environment and health authorities, economic analysis can be turned into a powerful tool for supporting their policies.

Why Use Economics?

First, economies can help to make explicit the benefits of environmental health improvements and the costs of the impacts. This provides additional arguments to encourage decision-makers to integrate environment and health considerations in their policies.

Second, current prices rarely reflect the full environment and health costs of the production or consumption of goods and services. Therefore, producers and customers have no economic reasons to reduce the impact they have on environment and health, as they do not pay prices that reflect this impact. Nor are they encouraged to take it into account in their investment decisions and lifestyle choices.

This could be corrected by reflecting as much as possible environment and health costs in the prices. Economic instruments, such as environmental taxes or tradable permits are a promising solution. A first step in that direction is the removal of subsidies that support practices harmful to the environment and health. In most of the cases, however, it would be difficult to remove distortive subsidies immediately and charge the full amount of environment and health costs. Nevertheless, negotiating plans and timetables to do so progressively, is a strong signal to the economic actors. It modifies their anticipation of future prices, as they know they will have to pay in the future for the environment and health costs they will create. This drives them increasingly to design their long-term choices and strategies in an environment-friendly way.

Finally, the setting of new economic instruments is usually under the responsibility of the Ministry of Finance. It also implies negotiations with economic sectors. Therefore environment and health authorities will need to play a more pro-active role in order to advance the integration of environmental health in sectoral and economic policies. Success will depend on their ability to discuss and present economic arguments in support of environmental health considerations.

A Promising Initiative

The present situation is that many environment and health authorities have few skills in using economic arguments and that economic sectors very often continue to ignore environment and health considerations.

International organisation—will also be invited to strengthen their co-operation in environment and health economics. In order to sustain the policy changes, promoting environment and health, co-operative efforts will aim.

- To support the development of the capacities of the environment and health authorities to use economic analysis;
- To improve the focus on health outcomes in national or inter-country processes dealing with environment and health issues. This will include the contribution of health expertise in these processes and the use of economic arguments to greater advantage;
- To exchange information early in the planning process of their respective programmes that use economic tools for addressing environment and health.
- To further co-ordinate their current and future activities in support of environment and health.

28

A Crucial Encounter

Genetic tests and treatments must not be allowed to create new forms of discrimination between those who, for whatever reason, can or want to take advantage of them, and those who cannot, mostly for lack of money. If a scientific discovery can form the basis of a technology, then it is highly probable that the technology will eventually be applied. Today this lesson of history is causing anxiety among politicians, scientists and public opinion concerned about the current far-reaching developments in biotechnologies.

It is now possible to penetrate to the very essence of living things as a result of spectacular scientific advances that are gradually revealing the innermost mechanisms of life. The technologies based on this field of knowledge offer humanity for the first time astonishing power to revolutionize the process of creating and developing human beings, and, ultimately, the human species. Technically speaking these breakthroughs could lead to the revival, in even more effective guises, of eugenic practices we hoped had been buried forever, Fortunately, this nightmare scenario seems highly unlikely.

But history also shows that new technologies are rarely applied without a framework of rules and procedures designed to ensure that they are beneficially used. Human progress has always been driven by the winds of freedom, including freedom of enquiry and initiative, but human beings have always tried to head in the right direction and to respect certain limits. The biologists have done their work; they have sown the seeds of vast possibilities. Now it is up to society to make sure that only the benefits are harvested. The biotechnology revolution beckons humanity to a crucial encounter between science and ethics.

Where human reproduction is concerned, as with technology in general, we must be guided by respect for three basic and interdependent principles dignity, freedom and solidarity.

For human dignity to be respected, each person must be regarded as unique. This position has far-reaching consequences for human procreation first of all, it rules out cloning as a means of reproduction because this technique, which is almost upon us, involves genetically "duplicating" an existing person. More generally, predetermining the basic characteristics of a future person, notably trying to enhance their future physical or mental capacities, violates the very essence of human individuality. This kind of engineering would end up by depriving individuals of that which is theirs alone—the mysterious processes whereby their unique genetic heritage emerges and interacts in its own unique way with their environment.

Advances in prenatal scanning and testing techniques may confront parents with grave new decisions. The danger is that various kinds of pressures or even regulations will develop which only allow "genetically correct" people to be born. This would be totally unacceptable. No authority—be it political, social or economic—should be able to enact such a "genetic order", still less impose it.

So increasing emphasis must be laid on solidarity. Genetic tests and treatments must not be allowed to create new forms of discrimination between those who, for whatever reason, can or want to take advantage of them, and those who cannot mostly for lack of money.

The risk of uncontrolled, unmonitored genetic engineering increasingly looms over us. But we are starting to see the emergence of a new "responsible" form of genetic engineering in which the power of science is subjected to the power of ethics an ethics that benefits everyone, not just a few, and looks towards future generations, not just short-term interests.

29

Sustainable Tourism and the Environment

Tourism is high on the international agenda. The 7th session of the Commission on Sustainable Development focused on tourism and subsequently work programmes on sustainable tourism are being developed. Also the Convention on Biological Diversity is embarking on tourism programmes and bilateral and multilateral financial institutions placed tourism high on their priority lists. The UN declared 2002 as the International Year of Ecotourism and the World Tourism Organisation adopted a Global Code of Ethics for Tourism at its General Assembly, held in Santiago de Chile.

The World Tourism Organisation forecasts that there will be 702 million international arrivals in the year 2002, that arrivals will top 1 billion in the year 2010 and that by 2020 international arrivals will reach 1.6 billion—nearly three times the number of international trips made in 1996, which was 592 million.

Travellers of the 21st century will go farther and farther. The Tourism 2020 Vision forecast predicts that by 2020 one out of every three trips will be a long haul journey to another region of the world. It is expected that China will become a major force in international tourism and the WTO predicts that about 100 million Chinese will take international trips by 2020, thus putting them in fourth place in numbers of travellers after Germany, Japan and the United States. By the same time, China will attract 137 million visitors—63.5 million overseas visitors travelled to China in 1998 and thus outrank France as the world's top destination. It is estimated that during 1999 France will receive a record number of tourists of more than 70 million; in 2007 France hopes to attract 90 million visitors. The key resource for the most popular tourist destinations is the most popular tourist destinations is the natural environment: coastal resorts, tropical rainforests, wildlife in national parks and alpine skiresorts, all rely on a mixture of natural beauty, good weather and safe condition to attract holiday destination is landscape and natural environment, followed by climate, the cost of the journey and the historical features of the place to visit, hence, conserving the ecological integrity and environment is imperative if tourism is to be sustained.

The pressure from millions of tourists on water and marine resources, on land and landscape, wildlife and habitat is enormous and often has devastating impact on the environment and the local population who are increasingly deprived of access to clean water and other natural resources.

In some regions, particularly in small island countries, tourism is one of the major reasons for wasting and polluting water: on average one tourist consumes at least 6 times more water than a local resident.

Major water wasters and polluters are golf courses. In many countries, golf has brought heavy ecological and social costs: deforestation, the destruction of bio-diversity and erosion; dispossession of peoples' homes and farms; over-consumption and pollution of water and very high use of pesticides and fertilisers which threaten local residents, workers, wildlife and the golfers themselves. A survey by the Japanese National Doctors Health Insurance Association has revealed that many golfers, caddies and residents living near a golf course suffer from skin inflammation, disorders of the ear, nose and throat and other respiratory illnesses to the inhalation of pesticides because up to 90 per cent of the chemicals sprayed on golf courses end up in the air. In some areas in Thailand, diseases emerged which, prior to the construction of golf courses, had not been known.

In some regions, golf courses have depleted water supply, agricultural production has come to a halt, peasants have become impoverished and forced to migrate to urban areas in search of employment. Golf courses take large amounts of land. It is estimated that each year world wide up to 5,000 hectares of forest are cut to clear land for golf courses.

Very often, the construction of golf courses forms an integral part of a comprehensive tourism project. Adjacent to the golf course condominiums and/or hotels are built, very often also a marina, an airport and a casino. Studies have shown that such a complex not only has touristic objective but is often connected to drug trafficking and money-laundering. Even the US State Department has emphasized the link between tourism, money-laundering and offshore banking.

Cruise ships are a major cause for pollution in the Caribbean, destroying maritime life and reefs by releasing waste into the ocean. Recently the Royal Caribbean, the world's second largest cruise line was fined a record sum of US$ 18 million of dumping waste oil and hazardous chemicals into the sea. The company admitted to routinely dumping wasted oil from its fleet and that it deliberately dumped in U.S. harbors and coastal areas many other types of pollutants, including hazardous chemicals from photo processing equipment, dry cleaning shops and printing presses. Some hazardous materials, including toxic solvents from dry cleaning operations, were illegally placed in the garbage aboard the ships. The material was then either incinerated on the ship or dumped in U.S. or foreign ports mixed with ordinary garbage.

It was announced that the Royal Caribbean Cruise reported a profit of US$ 338 million in 1997, a 93 per cent increase over the previous year, Carnival Corporation's Holland, the biggest cruise company with a turnover of US$ 3 billion in 1997 made a net profit of US $836 million, 25 per cent more than in 1996. Both cruise companies have recently been fined millions of dollars for dumping untreated bilge water, oil and other waste into Alaskan waters.

However, the impact of oil and hazardous waste on water, maritime life and coral reefs is devastating and all fines paid for the damage caused by the cruise ships will not revive dead corals.

A recent Green peace study on coral reefs—one of the marine world's great natural treasure—predicts that the coral bleaching which dramatically whitened many of the world's reefs last year will escalate rapidly under accepted global climate models and that the damage would wreak havoc in fisheries and tourism, disrupting the economies of many nations.

A WWF study recently published on "Climate Change and its Impacts on Tourism", warned that droughts, rising seas, flash floods, forest fires and diseases could turn profitable destinations into holiday horror stories. The report urges the tourist industry to persuade western industrialised governments to take more concerted action to reduce their nations' carbon dioxide emissions the main cause of global warming.

The Need for Action and Education

If government, the international community and the tourism industry want to save the world's major tourist destinations, immediate action is required. Governments and the tourism industry must abide to the principle that environmental protection is an integral part of tourism development. In order to protect the environment and mitigate the damages caused by tourism, some countries have decided to take action: The Spanish Island Minorca and the

Seychelles will introduce Eco-tax on tourism. This tax will be around US $ 12 per person in Minorca and its revenues are earmarked for the maintenance of national parks and the restoration of damaged coastline. Visitors to the Seychelles will have to buy a so-called "gold-card" at a price of 100 $ which entitles unlimited access to the country; income from this card will be used for sewage management and protection of fresh water supply.

Only if tourism investor and developers:

(a) consider the natural capacity for the regeneration and future productivity of natural resources.

(b) recognise the contribution that people and communities, customs and life styles make to the tourism experience and therefore accept that these people must have an equitable share in the economic benefits of tourism; and

(c) listen to local people in the tourist destinations, tourism may become sustainable.

Education and awareness raising campaigns at all levels are therefore imperative.

30

Pro-Poor Tourism

Opportunities for Sustainable Local Development

Tourism is the world's largest industry, with over 10 per cent of GDP globally directly related to tourism activities. Rising standards of living in the countries of the North, declining long-haul travel costs, increasing holiday entitlements, changing demographics and strong consumer demand for exotic international travel have resulted in significant tourism growth to developing countries. Tourism is the principal export for one-third of developing countries. Tourism brings relatively powerful consumers to Southern countries, potentially an important market for local entrepreneurs and an engine for local sustainable economic development. There is no reliable data on domestic tourism but it is growing rapidly in South America and in China and South East Asia; it represents a very significant economic opportunity for many local communities.

Tourism and Aid

Multilateral and bilateral aid agencies are wary of involving themselves in the tourism sector. In 1969 the World Bank created a Tourism Projects Department recognising that in the Mediterranean and Adriatic countries, and in Mexico, tourism had been a significant generator of foreign exchange and of direct and indirect employment, internationally in the late nineteen sixties, there was considerable concern about high rates of unemployment and the ability of developing countries to service debt. Tourism sector studies were completed in some 31 countries and tourism staff regularly participated in World Bank macro-economic missions—their reports focussed on the potential for growth in tax revenues, foreign exchange earnings and direct and indirect employment effects. The primary emphasis was on national economic impact. By 1978 when the World Bank closed its Tourism Projects Department of the Bank had provided loans and credits for 18 projects in 14 countries and it was the major source of funds and technical assistance for tourism development. The bank withdrew from tourism development for a range of reasons amongst which were anxieties about the role of the bank in funding projects to develop luxury hotels designed to attract wealthy travellers from the developed countries. This strategy was seen inconsistent with new policy objectives which prioritised the bottom 40 per cent, the Bank's priorities were shifting towards the poor, a group, which was gaining relatively little from tourism development. There was a growing literature that focussed on the negative economic, social and cultural impacts of unmanaged tourism on local communities. The fuel crises of the nineteen seventies also undermined some of the forecasts that had been made for the strength of the market and the Bank withdrew from the sector in parallel with most other multilateral and bilateral agencies.

The international agencies followed a macro-economic tourism agenda in the nineteen seventies and eighties focussing on tax and foreign exchange revenues at the national level, major hotel and resort development, international promotion and national and regional master planning all attracted funding. In the nineties the adoption of the new poverty elimination target of halving the number of people living on less than 1 US $ per day by 2015 refocussed development assistance on pro-poor growth. Multilateral and bilateral aid agency agendas are shifting towards micro economic strategies, which benefit local communities and in particular those below the poverty threshold. With poverty elimination now at the heart of decision aid, the potential for using tourism to generate pro-poor economic growth is being reassessed.

Since the mid-1980s, interest in 'green' tourism, eco-tourism and community tourism has grown rapidly among tour operators, policy makers, advocates and researchers. All of these focus on the need to ensure that tourism does not erode the environmental and cultural base on which it depends. The emphasis has been on minimising social, cultural and environmental impacts; rather than on positively affecting the livelihoods of the poor.

The Potential of Pro-Poor Tourism

There are a number of reasons to look again at tourism and to assess its potential to generate pro-poor growth, 80 per cent of the world's poor live in just 12 countries and tourism is significant or growing in all but one of them. Tourism is a very large sector, it is growing rapidly, and there is some evidence that it is relatively labour intensive. The consumer travels to the destination, creating additional—local—opportunities for the sale of additional goods and services—ranging from local pottery to a guided walk. Tourism can be used to diversify local economies; it can often be developed in remote and marginal areas with few other diversifications or export opportunities. These areas often attract tourists because of their high landscape, cultural and wildlife values. These natural resources and the local culture are amongst the few assets of the poor.

Pro-poor tourism generates net benefits for the poor. It can be defined as forms of tourism where the benefits to the poor are greater than costs which tourism brings them. Economic costs and benefits are clearly imported but social environmental and cultural costs and benefits are clearly important, but social and benefits also need to be taken into account. Pro-poor tourism aims to expand opportunities for those living on less than 1 US$ per day. Whilst it will also need to be sustainable preserving local culture, minimizing environmental impacts, it will be driven by the poverty agenda. Community-based tourism seeks to promote initiatives by local communities or individuals within them;

much has been learnt from these projects. Maximizing the poverty elimination effect requires that the emphasis is placed on involving those people who are living on less than 1 US$ per day and creating economic opportunities for them. Not all community tourism is pro-poor in this sense.

Effects on the Livelihoods of the Poor

Assessing the livelihood impacts of tourism is not simply a matter of counting jobs or wage income. Participatory poverty assessments demonstrate great variety in the priorities of the poor and factors affecting livelihood security and sustainability. Tourism can affect many of these, positively and negatively, often indirectly. It is important to assess these impacts and their distribution.

Tourism can generate four different types of local cash income generally involving different categories of people:

- wages from formal employment;
- earnings from selling goods, services, or casual labour (e.g., food, crafts, building materials, guide services);
- and profits arising from locally owned enterprises
- Income: this may include profits from a community run enterprise, dividends from a private sector partnership and land rental paid by an investor.

Waged employment can be sufficient to lift a household from insecure to secure. But it may only be an available to a minority, and not to the poor. Casual earnings per person may be very small, but much more widely spread and may be enough, for instance, to cover school fees for one or more children. Work as a tourist guide although casual, is often of high status and relatively well paid. There are relatively few examples of successful and sustainable collective income from tourism.

Negative economic impacts include inflation, dominance by outsiders in land markets and in-migration, which erodes economic opportunities for the local poor. Impacts differ between men and women. Women can be the first to suffer from loss of natural resources (e.g., access to fuel wood) and cultural/sexual exploitation, but may benefit most from physical infrastructure improvements (e.g. piped water or a grinding mill) where this is a by product of tourism.

Positive Development Impacts of Tourism

On the positive side, tourism can generate funds for investments in health, education and other assets, provide infrastructure, stimulate development of social capital, strengthen sustainable management of natural resources, and create a demand for improved assets (especially education). On the negative side, tourism can reduce local access to natural resources draw heavily upon local infrastructure, and disrupt social networks.

Tourism affects the livelihoods of the poor by changing their access to assets. In several cases, tourism's impact on people's access to natural resources or physical infrastructure has been identified as the most important benefit or concern.

Cultural Impacts of Tourism Can be Positive or Negative

Local residents often highlight the way tourism affects other livelihood goals--whether positively or negatively—such as cultural pride, a sense of control, good health, and reduced vulnerability. Socio-cultural intrusion by tourists is often cited as a negative impact. Certainly sexual exploitation particularly affects the poorest women, girls and young men. The poor themselves may view other types of cultural change as positive. Tourism can also increase the value attributed to minority cultures by national policy-makers. Overall, the cultural impacts of tourism are hard to disentangle from wider processes of development.

The overall balance of positive and negative livelihood impacts will vary enormously between situations, among people and over time, and particularly in the extent to which local priorities are able to influence the planning process. The application of a 'sustainable' livelihood framework is essential to developing pro-poor approaches. The distribution of livelihood impacts has to be considered. The poor are far from being a homogenous group. The positive and negative impacts of tourism will inevitably be distributed unevenly among poor groups, reflecting different patterns of assets, activities, opportunities and choices. The most substantial benefits, particularly jobs, may be concentrated among few. Net benefits are likely to be smallest, or negative, for the poorest.

Policies to Enhance Pro-Poor Tourism

Despite innumerable case studies of tourism development, there is relatively little assessment of practical experience in strategies to make tourism more pro-poor. Nevertheless, lessons can be drawn from a wealth of small initiatives (many from 'community tourism' or 'conservation and development' programmes), supplemented by expanding knowledge on 'pro-poor growth strategies', several policy implications clearly emerge.

1. *Put Poverty Issues on the Tourism Agenda*

A first step is to recognise that enhancing the poverty impacts of tourism is different from commercial, environmental or ethical concerns. PPT can be incorporated as an additional objective, but this requires pro-active and strategic intervention. There may well be trade-offs to make-for example between· attracting all-inclusive operators and maximising informal sector opportunities, or between faster growth through outside investment, and slower growth building on local capacity. These trade-offs need to be addressed.

2. *Enhance Economic Opportunities and a Wide Range of Impacts*

Two approaches need to be combined:

- Expand poor people's economic participation by addressing the barriers they face, and maximising a wide range of employment, self-employment and informal sector opportunities;

- Incorporate wider concerns of the poor into decision-making. Reducing competition for natural resources, minimising trade-offs with other livelihood activities, using tourism to create physical infrastructure that benefits the poor and addressing cultural disruption will often be particularly important.

3. *A Multi-level Approach*

Pro-poor interventions can and should be taken at three different levels:

- this is where pro-active practical partnerships can be developed between operators, residents, NGOs and local authorities, to maximise benefits;
- national policy level—policy reform may be needed on a range of tourism issues (planning, licensing, training) and non-tourism issues (land tenure, business incentives, infrastructure, land-use planning);
- International level—to encourage responsible consumer and business behaviour, and to enhance commercial codes of conduct.

4. *Work through Partnerships, Including Business and Tourists*

National and local governments, private enterprises, industry associations, NGOs, community organisations, consumers, and donors all have a role to play. It is particularly important to engage business, and to ensure that initiatives are commercially realistic and integrated into main stream operations. Private operators will not be able to devote substantial time and resources to developing pro-poor actions. NGOs and donors can help in reducing the transaction costs of changing commercial practice—for example, facilitating the training, organisation, and communication that would enable businesses to use more local suppliers. Changing the attitudes of tourists (at both international and national levels) is also essential if pro-poor tourism is to be commercially viable and sustainable.

5. *Incorporate Pro-poor Tourism Approaches into Mainstream Tourism*

Pro-poor tourism should not just be pursued in niche markets (such as eco-tourism or community tourism). It is even more important that mass tourism is developed in ways that benefit the poor. It is also important to assess which tourism segments are particularly relevant to poor. Domestic tourists are likely to be important customers.

6. *Reform Decision-Making Systems*

It is impossible to prescribe exactly how each tourism enterprise should develop in ways that best fit with livelihoods. The most important principle is to enhance the participation of the poor. Three different ways of doing this can be identified:

- Strengthen rights at local level (e.g., tenure over tourism assets), so that local people have market power and make their own decisions over developments.
- Develop more participatory planning.
- Use planning gain and other incentives to encourage private investors to enhance local benefits. These approaches require implementation capacity among governmental and non-governmental institutions within the destination, and require a supportive national policy framework.

It is time to reconsider the role of tourism in contributing to pro-poor development. Tourism should be judged against other possible strategies and where it offers the best opportunities for pro-poor growth, or where it can make a useful contribution by increasing the diversity of opportunities for the poor, tourism it should be considered. However, careful and effective local management will be essential if it is to contribute to meeting poverty targets and if tourism dependency is to be avoided.

31

Consumption Bomb

It is three decades since we passed the peak world population growth rate of 2.04 per cent. Annual additions too are now a decade past their peak of 86 million a year. They are currently running at 78 million a year and are heading downwards. A peak in total numbers, however, still lies at least four or five decades ahead. On the UN Population Division's 1998 projections, the total is likely to reach 8.9 billion in 2050. The long range medium projection, which has not been updated since 1996, expects world population to level out at just under 11 billion in 2200 AD.

However, this is based on assumptions that are increasingly questionable. More and more countries are reaching levels of female fertility that are not enough for replacement—below 2.1 children over the lifetime of each women. At the latest count there are 61 countries in this category. Of this 23 had very low fertility, below 1.5.

The situation is unprecedented in times of global peace on economic growth. The UN medium projection assumes that where fertility is very low it will rise again to 1.7-1.9 children per woman. In all countries where fertility is currently above replacement level of 2.1, it assumes that it will not fall below that level.

Yet fertility has fallen below replacement level in so many countries, which such different cultures and different stages of economic growth, that is increasingly looking as if low fertility may be here to stay. If this became the case, then world population may peak at some where between 8 and 9 billion. There after it may well begin to decline. The 1996 long range low projection has world population falling to 5.6 billion in 2100 AD.

None of this means that reproductive rights should have lower priority in future. Their contribution to the health and welfare of women and children are clear. Many poor countries in Africa and South Asia face huge population increases which will be hard to accommodate without major problems of land and water scarcity. In these areas reproductive rights receive a very high priority.

Increasingly our concern must focus on consumption, and how we can cope with the effects of its inexorable increase. Over the past 25 years world population increased by 53 per cent, but world consumption per person (Measured by income) by only 39 per cent. Assume that consumption per person will rise 100 per cent, while population will rise by only half that amount. As time goes on the preponderance of consumption will increase more and more.

There is a crucial difference between population and consumption aspirations. If fully assured of children's survival most people have quite modest desire for family size. But their desire to consume knows no upper bounds. As wealth increases, people double-up their possessions; two or three cars, two bathrooms, two rooms with all contents, two or three holidays a year.

Appliances improve every year and old ones "need" replacing. New needs are created that never existed before. Globalisation is making products cheaper than ever. TVs are no longer uncommon even in African shanty towns. The number of households is increasing as people live longer and family breakdown becomes more common. Smaller households consume considerably more per per cent than large. Moreover, consumption is politically very difficult to restrain. No one can get elected promising people they can earn and spend less, or re-elected if they fulfil their promises.

In view of this much of the burden of reducing our environmental impact will rest on technology. Technology will have to deliver major shifts in improving resource productivity, and in reducing the amount of waste we create. All our institutions and forms of management which affect technology will need to be geared to this end.

In some areas the record has been good and looks likely to remain so. Productivity has kept up with demand in the case of resources that are traded on markets, and that are under the direct control of people or companies affected by shortages or prices. Global food production has kept pace with demand: although land and cereal production per person has declined, average intakes of calories and protein have continued to improve and are at record levels. Malnutrition persists. but this is due to poverty and landlessness, not to the inability of the world to produce enough food. We have not encountered any limiting shortage of any key mineral resources or of energy. Nor are we likely to, because we continually economise and find substitutes, there has been a gradual reduction in the material used for each unit of production.

The prospects are much worse for resources that are not traded on markets or subject to sustainable management, as yet. These include groundwater, state forests, ocean fish, biodiversity in general. They include communal waste sinks like rivers, lakes and oceans, and the global atmosphere. In all of these areas it looks likely that things will get quite a lot worse before they get better.

These kinds of resources and sinks are not under the direct control of people affected by shortage or damage. People wishing to change the way a common resource or sink is used or managed have to pass through the legal or political system. They must organise, take out lawsuits against polluters. pressurise legislators and so on. Political responses are typically slow. Usually the majority of voters have to be convinced of the need for action before politicians will risk taking action. Even then powerful and rich vested interest will lobby hard for the status quo, and will often succeed in frustrating changes that are desired by a global majority. America's coal, oil, and car lobbies have stood in the way of any significant US commitment to reduce carbon dioxide output, and the US is the world's largest emitter of carbon dioxide.

Usually there has to be very wide spread and very visible environmental damage before action is taken. The thinning of the ozone layer fitted that category well and the response was swift. North Atlantic fishing reached that point in the 1990s, yet politicians shied away from taking adequate action until the last moment: fishing stocks plummeted and there was massive job loss. Global warming is still long way from the damage being widespread enough, and attributable clearly enough to human activities, for politicians to be ready to speed up the move into renewable energy.

The question with the common resources and sinks is always: will we react in time? The answer is all the more difficult because we usually don't know in advance what is "in time." Many critical changes are subject to threshold effects. When a certain point is crossed, very sudden and disastrous change can occur with little warning. In many cases we do not know where the thresholds lie.

Prudence dictates a preventive approach—a stitch in time saves nine. But the history of environmental problems shows that politicians rarely act decisively until the brink is reached, and it will always be touch and go whether we are pushed over it or not.

32

The Persistence of Indian Poverty and its Alleviation

Poverty has always been with us and for at least forty years its alleviation has been the professed objective of many strategies to improve the lot of the Indians. But the way in which it has been conceived, however, has been subject to considerable change. Relatively little attention was paid to the development of the poor themselves. Rather, they were portrayed as among the beneficiaries of development in larger systems which were to provide the dynamic force for the elimination of poverty (from the "outside" as it were). Development was principally something that happened to the poor—on a "trickle-down" basis.

The simple assumption that the poor would benefit from general economic growth, without paying any special attention to them, changed somewhat in the late 1950s, when it was perceived that the poor might not automatically benefit from macro-economic development, but that they must benefit if the social stability needed for overall economic growth was to be assured. From this point there emerged a specific line of antipoverty thinking to improve the income of poor people.

The manner in which the poor were to be integrated into the overall growth process, however, was very specific. It was concerned not so much with what the poor could offer to the growth process—as with what they should receive from that process. For all its merits the Basic Needs strategy, and the social "safety net" approach which followed it, basically emphasized the consumption needs of the poor—and not their surplus producing possibilities. On the contrary, a persistent theme in the discourse about the economics of the poor has been the need for some sort of transfer of resources to them from more productive and dynamic sectors of accumulation. In short, the poor have been portrayed as a net burden on the growth process.

It is possible to introduce an element of differentiation into this picture: given that it is rarely alleged that low wages are an obstacle to accumulation and growth, the poor who have been characterized as a burden have tended to be those not directly integrated into nascent large-scale systems of production: these are the poor "peripheral" to modern economic process—a group which encompasses a large proportion of the urban population in India (principally employed in the "informal" sector), as well as a vast mass of small, but relatively independent agricultural producers.

Implicitly, then, the concepts of "peripheral", small-scale and poor have been run together to form, in the realm of ideas, a more or less dependent mass. The number of people ostensibly in these categories is huge, and they seem to represent an enormous burden on development. They represent a development "problem", and an awesome one at that.

While substantial progress has been made in India in reducing the percentage of the rural population below nationally defined poverty line, the absolute number of the rural poor has increased. The growth of output did not bring about a significant improvement in the income share of the lowest nor an uniform reduction in the percentage of the rural population below the poverty line. The situation actually worsened. Less than half of the rural population in India has access to safe water or sanitation, and only 60 per cent had any access to health services. National data on life expectancy, infant mortality and literacy show improvements, but also the persistence of completely unacceptable conditions.

The pursuit of growth has not solved the development problem. Trickle-down has not worked or it has not worked enough. The massive persistence of poverty, particularly in rural areas represents a problem for the popular acceptance of continued economic adjustment; and it represents a problem for growth itself. The problem lies not only in the unintended consequences of the prevailing development paradigm, but in the viability of the paradigm itself. Part of the debt crisis arose from an inability to mobilize fully domestic assets, and from systematic resort to external resources. The unsustainability of this form of development has been amply demonstrated. Part of the answer to the challenge of development lies in a greater and more appropriate use of the resources of developing countries themselves.

A substantial part of these assets can be created by the poor who have been so marginal to past development efforts. The poverty of a nation and the poverty of people are not as easily separable as was often thought in the past. In many cases, it is difficult to envisage national growth without strong economic development among the poor themselves—not as objects, but as subjects of development.

The fact that this is insufficiently perceived is as much an expression of the development of social and economic interests as it is of the development or otherwise of economic theory. Development has frequently been associated with large-scale production and large-scale inputs of capital, and these new social and economic patterns have often defined development in their own image, i.e., in terms of the centrality of large-scale production and accumulation.

Poverty Alleviation

The perspective is not that growth achieved by the better-off will pull the poor out of poverty, but that the mobilisation and enhancement of the resources and activities of the poor themselves can uphold their dignity and free them from the shackles of misery, while at the same time making a vital contribution to overall sustainable growth.

Individually and collectively, the obstacles facing the poor are formidable. They are, however, not insuperable. Most of the forces creating poverty are essentially social. They reflect systems of resources allocation that are made by societies, and as such they can be reversed. Pricing policies, credit systems, and social and productive services, which neglect the poor, as well as gender discrimination, are not natural, universal and inevitable facts—and neither is the poverty they give rise to. One of the major obstacles to overcome in fighting poverty is the perception of poverty itself—and of the poor. In this regard, perhaps the most important point is that the poor are not idle, they work. Nobody is simply "poor". In other words, it is not just a state of being. In this regard, "poor" is more aptly used as an adjective rather than as a noun. The rural poor are poor farmers, poor herders and poor fishermen. In short, they are poor producers: their incomes are gained from their work. The answer to poverty lies in creating the conditions for them to earn more from their work. From this perspective, overcoming poverty does not mean less growth, it is a contributor to growth—for it means making the poor more productive. Too often in the past poverty alleviation has been seen as a burden on the economy, as involving a transfer of something for nothing in exchange. It need not be that way: it can be an investment in production, benefiting both poor and the national economy. Poverty has been defined as a production problem, and poverty alleviation as an investment.

Nobody wishes to be poor, and few accept it passively. The poor are rarely without initiative. What they lack are the means of pursuing it. In no small measure, overcoming poverty involves building upon this initiative and will, helping organize cooperation, and providing material support. This support does not have to take the form of handouts.

The problem of the poor is not that they cannot handle resources efficiently, but they do not have access to them.

The challenge of creating an institutional framework for credit for the poor is an expression of the general institutional challenge facing poverty alleviation: institutions are not oriented to the poor. Many factors enter into this, ranging from the costs of working with a large number of unorganized people, to the prevalence of myths about the improvidence of the poor, to a simple desire on the part of the better-off to monopolize scarce resources. The answer to this is to create institutional responsiveness, either through introducing demand-led organisation into existing institutions concerned with the poor or through promoting institutions created by the poor themselves. In both cases, participation by the poor is critical. The objective is not only to mobilize the individual initiatives of the poor, but also to mobilize their collective strength and capabilities. As individuals, many of the poor are virtually unreachable. As members of associations and groups they create their own channels for institutional access.

The poor as producers; the poor as credit-worthy handlers of material assistance; the poor as institutional actors—these are not elements of theory, but of practice and experience. Notwithstanding the growing acceptance of the need to do something "about" the poor, not everyone shares this understanding of poverty. As long as the poor are viewed from afar, the myths of poverty and the poor persist. Even those who over emphasize the need for social "safety nets" and handouts, while ostensibly helping the poor, maintain the image of helplessness, and of the need to do something "for" them. A closer view reveal something very different: tremendous work and initiative on the part of the poor, both based on their desire to do something for themselves. This is not a burden, it is an extraordinary social and economic asset. Again, viewed from a distance, poverty looks overwhelming. The closer view reveals very specific situations of opportunities and needs. These can be responded to—not only through soup kitchens, which should be seen as desirable in addressing emergencies only—but through strengthening the individual and collective means available to the poor to carve out their own path of independence and growth. The dynamics of poverty are reversible, but only in collaboration with the poor themselves.

Precisely because of past neglect of the poor as producers, a neglect involving a failure to involve them in the process of technological development, organisation, and capitalisation, the gap between the current and potential production of the poor is enormous. Investment in the poor is not a loss-making enterprise. Poverty is less a failure of the poor, than a failure of policy-makers to grasp their potential. Far from there being a tradeoff between poverty and growth, the persistence of poverty represents a limit to growth.

Mobilizing and enhancing the ability of the rural poor to expand their own income and contribute to national growth is not simply a process of raising incomes. It involves structural change in economies and societies. It involves helping the poor to position themselves securely within main-line economic processes. This means first increasing and improving their access to land—by land reform, land-titling, better management and better conservation, supported, where necessary, by irrigation, new technologies and improved infrastructure. Secondly, it means increasing the productivity and use of rural labour, emphasizing labour intensive technology and better training for new skills. Thirdly, it means making more capital available to the rural poor, mobilizing savings, providing infrastructure and developing financial services tailored to their situation and needs.

Not least, it mean acknowledging the important contribution of poor women in all of these areas of activity. At present, the contribution of women to the rural economy is seriously underestimated—the "invisible women" syndrome. Official statistics rarely make any effort to measure it, even though it is more than clear that not just unpaid household work but the farm and trading activities of women make a vital and significant contribution to the well-being of poor rural households. All the evidence suggests that the poorer the household the more hour's women work and the greater their investment in both economic production and family welfare. From a situation of multiple disadvantage as poor, as women and often as single parents, women can move to one in which they contribute and benefit three-fold—in the home, in society at large and, not least, in the development of the next generation.

Many of the measure that need to be taken to allow the poor to realize their potential do not involve more expenditures; they involve the elimination of economic distortions against the rural poor. These distortions have effectively taxed the poor, and mainly the rural poor, in favour of inappropriate and inefficient urban developments whose support has been at the root of widespread economic crises. To no small extent, helping the poor make their potential contribution to development involves no more than creating a "level playing field" and, when conceived in such a light, structural adjustment can make a vital contribution to both resumed growth and social equity. It is often felt that the poor are somehow "outside" the scope of national economic policies. This is virtually never the case. They are affected by national economic policies, but this inclusion takes a very special form: exposure to the costs, and exclusion from the benefits. In this regard, there is a certain irony in the view that small-scale producers "need" subsidies to survive. In fact, it has been the development of large-scale production in agriculture (and industry) in India that has been heavily dependent upon subsidisation over the decades—benefits, which smaller-scale producers have rarely enjoyed.

This is characteristic of many forms of large-scale production in India—although nominally at the cutting edge of efficiency and productivity; it is they rather than the small-scale producers who have been dependent upon transfers and protection for their reproduction.

Change in the environment of poverty necessitates greater awareness of the root causes of poverty on the part of policy-makers. However, the realisation of the social and economic potential of the rural poor is not just a question of economic policy and investment. It also involves the development of a general social framework in which the economic and social interests of the poor can be freely articulated and responded to. It means instilling democratic and participatory values at every level in society and not just at the level of nationwide institutions. The most valid spokesmen of the poor are the poor themselves.

The opening of economic and social opportunities to the poor offers the possibility of more stable and sustainable change. The alternative is for societies to polarize further, for the welfare burden to grow to greater proportions and for a widening gap to develop between the modern and traditional sectors. In the end, the continued poverty of the rural areas will be a brake on the output of the advanced sector, eroding the potential for self-sustaining growth.

33

Employment and Poverty Alleviation

Today the key socio-economic problem is large-scale unemployment. Spreading joblessness brings many other problems in its wake. It erodes national incomes and living standards, aggravating the already grindingly difficult job of promoting development and alleviating poverty. Joblessness also raises government budget deficits, increasing macro-economic instability while soaking up investment for productive capital expenditure, education, training and relief aid. And joblessness ruins lives and communities by depriving people of the dignity and satisfaction that comes with earning one's keep and making a contribution to the well-being of family and society.

Theories about how best to nurture development (and thus create jobs) have shifted considerably over the last decade. The state role has evolved, in the minds of many, from being a source of relief for the problems of unemployment, poverty and underdevelopment, to being a fundamental cause of these problems through the distorting impact of its intervention on the market.

However, the more market-oriented philosophy that grew up during the 1990s has yet to provide convincing solutions in practice at least not on a grand scale and especially not in terms of job creation as the present jobless economic recovery demonstrates.

The weakness of the current recovery and past approaches to economic development can be traced to the failure to consider employment as the predominant means of promoting growth and alleviating poverty. In policy circles it has too long been an almost ignored priority.

Current trends thus bode poorly, particularly as unemployment rates soar. In light of the circumstances, we need to begin re-examining some of the fundamental questions—if only to find out what has gone wrong with the answers.

Minimum Wage?

Let's begin with wages. With corporate restructuring in full force on a global scale, are low wage rates required to raise employment and maximize profits? A top manager of a multinational consumer electronics group certainly thinks so; he likened the perfect factory to a ship "so that we could move it around the world to where labour was cheapest". Perhaps, but this bottom-line emphasis on unit labour costs ignores at least two other factors; namely, that higher wages can act as a screen to select more productive workers and that higher wages translate into better productivity via improved worker nutrition, increased consumption and a generally healthier quality of life.

If higher wages being these benefits (and it is an open question) should government insist that there be a minimum wage rate? Neo-classical economists tend to respond "no", assuming that a higher wage rate puts money into the pockets of some low wage workers while forcing many others out of work because companies cannot afford to pay them.

Technology Transfer

The impact of technology is another area in need of study. Technological innovation is usually labour-saving and tends to originate in industrialized countries, moving toward developing countries like India, Pakistan where labour tends to be low cost and abundant. Would it therefore make sense to slow down or somehow restrict technology transfer, especially to development markets, in the interest of preserving employment?

The answer here is clearly—no, historical evidence abundantly demonstrates that attempts to retard technological progress bring about greater poverty and lower growth. Technology, infact, is at the heart of the new endogenous growth theory which is very much in vogue among development economists today. Slowing down or inhibiting technology transfer would certainly dash many countries' development hopes and aggravate poverty. However, the relationship between technology, development, employment and poverty alleviation is not without its complications.

In the 1980s, the buzz word among development specialists was "appropriate technology", i.e., small-scale and labour-intensive technologies that would increase productive output while allowing an equilibrium solution to be found such that the ratio of the productivity of labour to that of capital is proportional to their relative prices. The conditions for this "small is beautiful" approach to technology tended to be best met in agricultural production. However, where manufacturing industry is concerned, the small-is-beautiful approach foundered badly when the only viable technological alternatives proved to be highly capital-intensive.

Development Gap

A wide gap has emerged between developing countries with an inward focus (which tended to be protectionist and pursue policies of import substitution) and those with an outward focus and a policy of pursuing export-led growth. Competing in international markets requires technology that is as good as or better than that found in advanced, industrialized nations. Small, therefore, is not beautiful in the global manufacturing economy where product standards are high and the elasticity of substitution between labour and capital is very limited.

The drive to obtain state-of-the-art technology thus leads to a policy conundrum: it is a pre-condition for success in manufactured exports, but the impulse to compete successfully in this most lucrative sector speeds up the transfer of technology from the developed to the developing world, thus reinforcing the bias toward labour saving equipment in developing countries and accelerating a process that is seen as a source of job loss in the industrialized countries.

Technology and Jobs

Before concluding that modern technology transfer is inimical to employment in developing countries, we have to distinguish clearly between technology's static and dynamic consequences. In a static sense, it is true that highly capital-intensive export industries may not create much employment on a net basis, but the dynamic effects of technology transfer do contribute to economic growth. And growth, in turn, generates multiplier effects in the form of demand, which stimulates ancillary production activities (like food processing or consumer goods) that rely on more labour-intensive technologies.

The problem is that the diffusion and application of technology on a global scale blurs the categories of international product specialisation and creates a much more competitive and conflict-prone international environment.

For example, we have already seen the Asian Tigers move from producing goods such as textiles and processed food to producing hi-tech and value-added consumer durables. This advance is only possible due to the growth of human capital (facilitated by investment and higher incomes) and it leaves production of textiles to other industrializing countries, like Indonesia, the Philippines and now China. But the dynamic comes at the expense of jobs in industrialized regions, like the US and the EC, which lost more than a quarter of their work force in textiles during the 1980s. In spite of job losses, advanced countries continue to produce textiles, notwithstanding major differences in the hourly wage rates for spinning and weaving and the fact that essentially the same hi-tech equipment is being used in most production centres.

Protectionism

What has happened in textiles is happening in other industrial sectors (automobiles, for example) as well. The intense market competition is providing to be a source of trade conflicts, and possibly protectionism, as jobs come under increasing pressure.

For many workers and managers, the benefits of foreign direct investment look increasingly like a zero-sum game for employment, and there is a real risk that the tenuous link between overall growth and employment will break down altogether. It is hardly surprising that we are already seeing negatively affected workers and local businesses clamouring for protection in advanced countries.

Governments Role

The concerned governments are suppose to carry out much of this research. The three initial lines of inquiry follow from three reasonable assumptions about the future.

- First, increase in welfare and consumption subsidies are out; investments in training and human capital are in. How can investments in human capital be directed to positive employment effects? Is it perhaps not time to explore more fully benefit schemes targeting the unemployed and the unskilled poor providing them with the type of subsidies that would enhance their human capital, improve their health and productivity through better nutrition and preventive medicine, and restore the dignity of holding a job?
- Second, given the quasi-inevitability of increased automation in manufacturing, how can other sectors (particularly agriculture and services) be developed to

export their long-term potential for employment creation?

- Third, given the inevitable pressures of work and productivity in the global economy, what sort of alternative institutional arrangements need to evolve with respect to industrial relations, employment and work conditions?

Finding answers to these and other questions will require no small amount of new thinking, but parochialism or a failure of imagination would be fatal flaws in this global era.

34

Food Production

During the last 25 year, world agriculture successfully expanded food production faster than population growth. This can continue for the next 25 years and beyond, if appropriate action is taken. Although world food stocks are currently low and grain prices high, the world is not about to run out of food. We can produce enough food for future generations if we choose to do so.

The widespread food insecurity, unhealthy living conditions, and abject and absolute poverty in many developing countries are already threatening global stability. Failure to assure sustainable food security will foster the very conditions that will further destabilize and polarize the world in the years to come with tremendous consequences for all people.

The Basic Facts

Poverty is widespread in developing countries, with over 1.1 billion people living on a dollar a day or less per person. Human resource development in developing countries is lagging: 1 billion people lack access to health services, 1.3 billion do not have access to adequate sanitation systems, and one-third of primary school enrolls drop out by Grade 4. Natural resources, upon which future food production depends, are being degraded at alarming rates: almost 2 billion hectares of land have been degraded in the past 50 years: about 180 million hectares of forests have been converted to other uses during the 1980s, marine fisheries are collapsing around the world, and regional and seasonal water shortage afflict many developing countries. Improved appropriate technology is essential to increase productivity. Yet low-income food deficit developing countries are grossly under-investing in agricultural research and many are reducing their support.

It calls for sustained action six priority areas. First, we must selectively strengthen the capacity of developing country government to perform appropriate functions such as establishing or clarifying property rights, promoting private-sector competition in agricultural markets, and maintaining appropriate macro economic environments. Predictability, transparency and continuity in policy making and enforcement must be pursued.

Investing in People

Second, we must invest more in poor people in order to enhance their productivity, health, and nutrition. It is not only unethical but economically wasteful that a large share of the World's population is malnourished, illiterate, sick, and without access to productive resources. Access to primary education, primary health care, reproductive care and family planning information, and clean water and sanitation must be assured for all people. Access by the poor to productive resources and remunerative employment must be improved. Empowerment of women must be supported.

Third, we must accelerate agricultural productivity. Agriculture is the life-blood of the economy in low-income developing countries. In those countries, it provides up to three-quarters of all employment and half of all incomes. There are very strong links between agricultural productivity increases and broad-based economic growth in the rest of the economy. Agriculture is an engine of growth in low-income developing countries. National and international agricultural research systems must be mobilised to develop improved technologies focused on developing countries, and extension systems must be strengthened to disseminate the improved technologies and techniques. Low-income countries currently spend less than 0.5 per cent of the value of agricultural production on agricultural research compared to 2 per cent spent on agricultural research in middle and high-income countries. An increase of agricultural research expenditures in low-income countries to at least 1 per cent of the value of an agricultural output is urgently needed, with a longer term target of 2 per cent. National agricultural research must be supported by a vibrant international agricultural research system that undertakes research with large international benefits applicable across boundaries. Current investments in international agricultural research are grossly inadequate to provide the support needed by developing countries. It is of critical importance that agricultural research result in reduced unit costs of production. Such cost reductions will make food economically accessible to low-income consumers, and permit producer incomes to increase. To assure relevance of research and appropriate distribution of responsibilities, interactions between public sector agricultural research systems, farmers, private enterprises, and NGOs must be strengthened.

Fourth, we must assure sustainability in agricultural production and sound management of natural resources. Farmers, local communities, and governments must be encouraged to establish and enforce systems of rights to use and manage natural resources, to improve the way water is allocated and used, to reverse land degradation where it has occurred, to reduce the use of chemical pesticides and promote integrated pest management programmes, and to implement integrated soil fertility programmes in areas with low soil fertility. Local control over natural resources must be strengthened and local capacity for organisation and management improved. Investments in less-favoured geographical areas, that is, areas with agricultural potential, irregular rainfall patterns, and fragile soils must be expanded. Most poor people in developing countries reside in rural areas, and most rural poor reside in less-favoured areas. Yet, most investments, including agricultural research investments, still focus on the more-favoured areas. If we are serious about reducing poverty and protecting the natural resource base, the balance between the less-favoured and more-favoured areas must be redressed.

Fifth, we must reduce food-marketing costs in low-income developing countries. The cost of bringing food from the producer to the consumer is very high in many of these countries. Efficient, effective, and low-cost agricultural markets must be developed in order to bring these costs down. Inefficient state-run firms in agricultural in-put markets must be phased out; investment in developing and maintaining infrastructure, especially in rural areas, must be forthcoming; policies and institutions that favour large-scale, capital-intensive market agents over small-scale, labour-intensive ones must be removed; development of small-scale credit and savings institutions must be facilitated, and technical assistance to create or strengthen small-scale, labour-intensive competitive rural enterprises must be provided.

Sixth, we must expand and realign international development assistance. Many years ago, industrialised countries had agreed to allocate at least 0.7 per cent of the Gross National Product (GNP) to international assistance. Most countries have not reached or do not maintain this target. Not only must the industrialised countries increase international development assistance to reach the 0.7 per cent target, but they must realign it to low-income developing countries. Also contrary to the middle- and higher-income developing countries, the poorest countries are not able to gain access to capital from the rapidly expanding international commercial capital market. Developing countries in turn must seek measures to diversify sources of external funding, stem capital flight; and improve the effectiveness of the aid they receive.

35

No Progress without a Secular Society

Every day, women continue to be victims of rape, trafficking, acid-throwing, dowry deaths and other kinds of torture. At the opening of this new century, women are still not considered as equal human beings in many parts of the world. Religion and patriarchy continue to have an all-encroaching hold on their lives, maintaining and justifying their age-old oppression. In some South Asian Societies, this hold is even increasing.

I do not believe that there can be real equality in a society dominated by religion. Western countries speak repeatedly about the necessity of economic development to alleviate poverty. But this is not enough. Some oil rich countries may be economically developed, but women are deprived of all rights. The supremacy of religion is incompatible with freedom of expression, women's rights and democracy. This is why I see religion as the main enemy of women's development.

We have to act on several fronts at once. First of all, improving access to education. In a society like Bangladesh, 80 per cent of women are illiterate. For centuries women have been taught they are the slaves of men. It is very hard to change their minds, to make them aware of their oppression, to give them a sense of their independence. This educational effort has to go hand in hand with a secular feminist movement in society. Such movements have to start within the country and they cannot take hold when people are uneducated and unaware of their oppression. I'm not sure you can accomplish much from the outside, except to expose in the media the atrocities women in all too many countries face in their day to day lives.

In some countries, this movement is emerging, but very timidly, and it has a slim margin of maneuver. It has the uphill task of fighting for the repeal of religious laws and the introduction of a uniform civil code. So far, it tends to be constituted by a few individual feminists who are forced to be diplomatic, to compromise with fundamentalists, be they men or women. But they are trying to change the system, step by step, and it will take a very long time. People are not yet ready to do away with religious laws that impact upon every aspect of society, from education and health to the workplace and the home.

For women's status to change, we also need enlightened leaders who believe in equality. In countries of South Asia women with a strong voice do not have the support of political leaders, whether they be men or women. Look at the countries in which women are in politics, or even heads of state. Does it follow that women in those countries are emancipated? Because of long-standing vested interests, such leaders continue to back measures that oppress women. They are not ideologically committed to changing these conditions. In South Asia, most of the women who become heads of state are religious, and like men, they adhere to the religious objectives of the establishment. Until a society is not based on religion and women and considered equal to men before the law, I do not think that politics will advance the cause of women.

Until a society is not based on religion and women are considered equal to men before the law, I do not think that politics will advance the cause of women. In Western countries, women are educated, they are treated equally, they have access to jobs. In these conditions, their participation in politics has a meaning.

Education, a secular feminist movement and leaders—both men and women—committed to equality and justice. This is what it will take to change the dire conditions which too many women still face today. It will take a very long time, but we are here to work towards that end.

36

What's Driving Migration

The scale and diversity of today's migrations are beyond any previous experience. Rapid urban growth and environmental degradation in rural areas have led to internal migration affecting hundreds of millions of people. Migration is now seen as a priority issue equal in political weight to other major global challenges such as the environment, population growth and economic imbalances between regions.

Families and households form the basis for economic growth, social development and personal fulfillment. Decisions, by individual women and men on marriage, family, a place to live, shape the destinies of communities and nations. National policies and international conditions provide the context for individual decision-making. Effective development policies, including population, reproductive health and family planning policies, address this reality.

Data on national and global population trends set the agenda for national policy. An important element of population programmes is gathering data that will allow policy-making responsive to the realities of daily life, and to the needs and aspirations of individuals.

The dominant feature of global demographics is still growth. Age distribution is a growing concern, as the numbers of young and elderly people, grow, relative to the working-age population. The world is growing steadily more urban. From being a sign of strength and dynamism in the national economy, the rate and scale of urban growth has become increasingly a cause for concern. The influx of migrants to the biggest cities may be weakening both urban and rural sectors.

International migration is small in extent compared with internal movements, but has a disproportionate impact. Both internal and international migration are driven by population growth, and by inequities between countries. Migration is one of the choices which shape people's lives and the destiny of nations. But it can also be a symptom of inequity and underdevelopment. Migrants are by definition the most vulnerable members of the host community. Their living and working conditions should be protected.

Open and frank exchange of information and views between host and sending countries is needed more than ever. The aim of the international community should be to protect the right to move, but to ensure that movement is voluntary and that it stimulates rather than holds back personal and national development. "The point of departure should be the human right to live and work where one pleases, so long as it does not infringe on other people's rights to do the same."

The Urban Transformation

The rural sector is declining in importance and its contribution to national economies. It is increasingly part of a unified economy based on the city. Contact with the urban areas is easier than ever and is encouraged by rural development.

Temporary and circular migration is giving way to more permanent settlement. The largest cities are under increasing strain, and residents are encountering increasing difficulties in improving or even maintaining living conditions. Nevertheless, migration continues, driven by a variety of forces both positive and negative. The choice to move can be part of a strategy for survival or personal development; but it is often enforced by external conditions.

The urban transformation is irreversible, but the rural sectors must also be strengthened to balance the developing economy. Attention to gender issues will be crucial in ensuring a successful transition. The forces driving internal and international migration have much in common. Demographic pressures are contributing to both. As the pressures encouraging migration increase, the options for migrants become more limited. This collision is contributing to the atmosphere of crisis surrounding both urban and international migration.

Costs and Benefits

Migration is the result of individual or family decisions. But it is also part of social process. In economic terms, migration is as much a global phenomenon as trade in commodities or manufactured goods. It is part of a broader pattern, and evidence of changing economic, social and cultural relationships.

But migration may be evidence of a different kind of relationship; the combination of poverty, rapid population growth and environmental damage is a powerful destabilizing factor driving urban growth and eventually international migration. On the recipient side, migration has usually been seen as evidence of a thriving economy; today's industrial states were built in part by migrant labour, skills and investment. In today's increasingly uncertain conditions, migration may be seen as a threat to the security and well-being of the local workforce and society at large.

The only effective means to reduce migration pressures over the long term are to slow population growth; to stimulate economic growth and job creation at home, and promote the development of the individual and the family as the basic economic and social unit.

A Question of Gender

It is often assumed that most migrants are men, in reality, women make up nearly half of the international migrant population. Gender differences in social and economic roles affect migration decision making, household strategy, and the sex composition of labour migration. Attention to the gender dimension of migratory movements ought to be an important component in population and development planning.

Women frequently take the initiative in migration decisions, which may reflect limited opportunities in rural areas. Low status limits women's choices at home and may increase pressure to migrate, but it may also affect life in the host community. Opportunities may be limited by lack of education or skills, or by customer limitation on women's freedom of action outside the family or ethnic group. Paid employment for migrant women is usually in the lowest wage, least secure, and lowest status jobs, mostly in housework, child care and trade.

Most educated women end up in the same low-status, low-wage production and servicejobs as unskilled female migrants. Men too, experience downward mobility, but the contrast in the decline in women's employment status is far greater. Despite these disadvantages women migrants have become significant economic actors. Their status may be improved by migration, but the advantages are not clear-cut. Women's status as migrants is affected by their vulnerability, and by their lack of reproductive freedom. To ensure improved status they will need both legal protection and essential services, including reproductive health services.

Refugees

Refugees in the 1990s are overwhelmingly in Asia, Africa and Latin America. Their numbers are large, about 17 million, and growing rapidly. A further 3.5 to 4 million were thought to be in "refugee-like situation", though estimates are probably extremely conservative, and an estimated 23 million people internally displaced.

It is important to recognize the common roots of refugees and other forms of mass movement of populations.

At the same time, despite the difficulty of distinguishing between political and socio-economic causes of migration, there is a clear need to distinguish between refugees and other groups of migrants. Participation in international efforts of burden-sharing would ensure that most refugee problems would be dealt with in their regions of origin.

Conclusions and Policies

Migration highlights linkages and interdependencies within countries, with many implications for development agendas, including population programmes and development assistance.

Policies to regulate or moderate international migration have concentrated largely on urban growth. They have been only intermittently effective. The most successful have concentrated on stimulating rural development and the growth of alternative urban centres

Migration is also a personal or family decision, which is affected by external conditions such as poverty or environmental degradation, improving conditions of personal and family life can make a crucial difference in the decision to migrate, reducing dependence on migration as a strategy. Because migration is the result of personal and family decisions, it can be influenced by policies that improve the quality of life.

This offers the opportunity for policies emphasizing individual development, among them education, health (including reproductive health) and family planning. Such policies are particularly relevant to the strategies must take into account gender differences in social and economic life and the differential effects of policies.

Migration decisions are about family security and long-term-life-chances, rather than simply the maximisation of income. They are ultimately strategies designed to look after the individual's and the household's needs, safeguard their security, and respond to their aspirations. If the goal is to reduce migration pressures through development it will be essential to increase the capacity but reduce the need to migrate. Long-term external support will be required to make such policies a reality, particularly in areas of rapid population growth and potential mass outward flows. Highly co-ordinated allocation of development assistance can be help establish priorities and focus attention on basic needs. The challenge to both international donors and co-operating governments is to direct programme spending to the areas where it can be most effective.

37

Major Cyclones in Andhra Pradesh

Some Observations

Cyclone is considered as the most devastating of all natural phenomena. Repeat cyclones create no sense of security to the wealth and life of the people. It is also well-known that the sudden and rapidly developing cyclonic storm not only disrupts the prevailing order of life and produces danger, death and loss of property to large number of people residing within a geographical area but also creates environmental imbalances.

Andhra Pradesh is one of the states facing moderate to severe and repeated cyclones affecting the life of the people of the state and especially coastal districts. In the history of Andhra Pradesh, there is a long list of cyclones. The cyclone which hit the town of Machilipatnam in the year 1864 was one of the severest in the history of the world. The tidal wave was so vast and rushed with such force that it submerged the coastal areas deep up to twenty miles. It took many weeks for the flood water to sink. The entire area, which was submerged under seawater, became saltish and consequently infertile for many years to come. It appears that in the year 1796 also Machilipatnam and surrounding areas were hit with a big tidal wave which claimed over 20,000 people besides loss of property worth more than Rs. 250 crores then. The impact of cyclone of 1864 was more on life and property of the people. More than 35,000 people and 3 lakh livestock dead and worth o nearly Rs. 600 crores property was lost. Almost all the houses at Machilipatnam were completely ruined. Even the mighty fort of Britishers turned into shambles but for one building and the Bell tower of the church. With this calamity and Machilipatnam being prone to frequent cyclones, the Britishers shifted all their establishments to Madras, after this cyclone. The Britishers erected two monuments in memory of the 1864 cyclone victims—one of Robertson Square in the heart of the town and the other at Machilipatnam Fort.

Relief measures were taken up to persons who were deprived of the essential needs of life because of natural disaster resulting from cyclone of 1864. Cyclone relief consists largely of emergency provisions of food, clothing, medical care and shelter with the aid of voluntary contribution from other communities and countries. Even International Red Cross and other voluntary organisations did not take cyclone relief as one of the chief activities. Only in 20th century these organisations have been extending assistance to the victims besides greater role of the central and state governments in relieving the population affected by the cyclones.

The second major cyclone in the state like that of 1864 cyclone, hit the coastal region on the night of 19th and 20th November, 1977. There was terrific gale with speed ranging 120-150 km. per hour. Coastal districts of Andhra Pradesh such as Krishna, Guntur, East Godavari, West Godavari and Prakasam and to a small extent Srikakulam, Nellore and Visakhapatnam districts were affected. The most affected areas were Divi, Machilipatnam, Bapatla, Repalle and to some extent in Chirala Taluk. Huge tidal waves engulfed the coastal region of the Divi, Machilipatam and Repalle Taluks. There was extensive damage to life and property. Approximately 71 lakh persons in 2302 villages were affected by the loss of property, crop and other damages. 7932 persons lost their lives. In Krishna district alone 6706 persons died. The loss of cattle was 2,32,046 and 45,53(

other livestock. There was substantial damage to houses—86,650 huts were completely washed away. Approximately 2,16,000 persons have either lost their houses or have suffered other loss. There has been widespread damage to crops worth of Rs. 450 crores and substantial damage to public buildings such as roads, electrical installations, highways, railway, telephonic installations, irrigation canals, etc.

To overcome the effects of 1977 cyclone, 172 relief camps were opened in Krishna, Guntur, East Godavari and Prakasam districts. About 2 lakh persons were provided shelter immediately. More than 25,000 quintals of rice and food packets were distributed. Medical special staff was sent to the affected areas to assist the Collectors to take measures against out-break of epidemics. Financial assistance was given for house repairs and the Government sanctioned remission of land revenue in the most affected areas of Machilipatnam, Repalle, Bapatla and Divi taluks.

Andhra Pradesh again experienced a major cyclone disaster in May 1979. Besides loss of life there was enormous damage to public and private properties in coastal districts. Prakasam, Nellore and Kurnool districts were the worst hit. The heavy rain under the influence of the cyclone, affected the districts of Guntur, Krishna, West and East Godavari on the coastline and Kurnool and Cuddapah districts of Rayalaseema and Mahaboobnagar district of Telengana. This was most unusual occurring in the month of May and again touching all the three regions in the state. Due to precautionary measures like providing information in time to the people of the state evacuating all the people to safer places mobilising and gearing up of administrative set up to take immediate steps, death toll and loss of property were greatly reduced though the intensity of the cyclone was double than that of the one which occurred in 1977.

Nearly Rs. 60 crores worth of crops were damaged in the State. The crops in Nellore district were heavily damaged followed by East Godavari, Guntur, Cuddapah and Prakasam districts. Due to intensive precautionary measures during precyclone period, the death toll was drastically kept under limit with just 750 human lives though the severity of it was very high compared to the earlier cyclones. About 3 lakh livestock were lost due to improper care taken during pre-cyclone and during cyclone periods. Out of the six cyclone affected districts, Prakasam lost heavily its livestock. Most of the deaths in this district were due to surging waters, which breached the tanks with extensive loss of livestock and damage to the property.

More than cyclone relief measures, 1979 cyclone reveals that the success of the excellent preparedness measures taken by the State could reduce the death toll greatly. Wholesale evacuation of the people living in low lying areas in coastal areas prevented not only the number of deaths and loss of property but also could reduce the risk of undertaking relief measures.

Another severe cyclone was experienced by the people of Andhra Pradesh in October, 1983. As Andhra Pradesh is called the rice bowl of India and more especially East and West Godavari districts, paddy crop was completely damged. All the major rice producing districts such as East Godavari, West Godavari, Nizamabad and Karimnagar districts were severely affected besides Visakhapatnam, Khammam, Warrangal and other districts, though the loss of human life was below 200, the cyclone could affect 50 lakh people in 6,322 villages of 14 districts. More than 2 lakh houses were completely damaged and nearly 1,50,000 houses were affected partially; 12,000 cattle and 31,000 livestock perished in the cyclone. Standing crops in 1,38,500 hectares were completely lost and crops in about 31,13,150 hectares were partially damaged in addition to dry crops in 3 lakh hectares of land. The loss of both public and private properties was estimated around Rs. 600 crores. The road length of more than 8,500 kilometres damaged completely. This was considered as one of the most severe cyclones in recent years as it could affect the economic structure of the State very badly.

Like that of 1979 cyclone, with utmost pre-cyclone preparation and protection, death toll and loss of property were minimised to the lowest possible level. Thanks to the immediate step taken by Telugu Desam Government and officials incharge in the districts, a major threat was averted with minimum loss by evacuating the people to safer places in almost all the coastal districts. Indian Army and Navy, Police and other officials could evacuate more than 50,000 persons in Visakhapatnam, East and West Godavari, Guntur, Krishna, Nizamabad, Karimnagar and Nalgonda districts to safer places. Hon'ble Chief Minister, Sri N.T. Rama Rao, has sanctioned immediately Rs. 30 crores towards relief without waiting for Central assistance.

After five months, in February 1984, the districts of Nellore and Chittor have again affected by a cyclone. The most unusual cyclone in the month of February could bring misery to farmers in more than 150 villages by destroying standing paddy crop and gardens worth of more than Rs. 20 crores and loss of property in the form of collapse of houses worth more than 50 crores. It is, however, the administration of organised emergency relief and rehabilitation could set right the imbalance. The Cyclone of 1996 has the same impact on A.P. economy.

The cyclones which affected the people of Andhra Pradesh demonstrate that the people affected by cyclone disaster are not confined to the immediate geographic areas of destruction, death or injury. All persons who are related to, or who identify themselves with, persons and organisations in the stricken area are also affected. Thus the effectiveness of cyclone relief measures are usually national in scope. In each cyclone period, effective preparations were made to face cyclone with necessary arrangements by evacuating the population to safe places.

But there remains a problem of co-ordination and control in undertaking relief operations. Shortly after the cyclone impact, thousands of persons have to converge on the affected area and on first-aid stations, hospitals, relief centres and communication centres near the area. Along with this movement of persons, incoming messages of anxious inquiry and offers of help from all parts of the country and even from foreign countries should come forward to set right communication facilities and to supply food, clothing, bedding and other materials. This action should continue for a few weeks following a cyclone. At times of cyclone generally there would be confusion due to lack of systematic procedures for maintaining a central strategic overview of the cyclone to apply controls and resources where they are most critically needed. Communication is often inadequate partly because of the destruction of communication facilities but more generally because of improper use of these facilities, with greater coordination and cooperation from the people and making them feel a great sense of urgency to help the victims, these effects of the cyclones would be greatly minimized in future.

Bibliography

Anand, R.P., *Legal Regime of Sea Bed and the Developing Countries,* 1975.

Bhatt, S., *Environment Protection and International Law,* Radiant Publication, Kalkaji, New Delhi, 1985, pp. 122.

Bhatt, S., *Environmental Laws and Water Resources Management,* Radiant Publication, India, and Advent Books Inc. New York, 1986, pp. 355.

Behrman, Danial, In *Partnership with Nature—UNESCO and the Environment* (Paris, 1973).

Bell, Daniel, "Technology, Nature and Society", *American Scholar,* Summer 1973.

Bentley, Glass, "Biology and Human Values", USIS, New Delhi.

Book of Nature. The Way Things Work, published by George Allen and Unwin Ltd., 1981, pp. 525.

Boulding, Kenneth E., "New Goals for Society", S.H. Schun, ed., *Energy, Economic Growth and the Environment.*

Carr, E.H., *What is History,* 1961.

Darlington, C.D., *The Evolution of Man and Society* (London, 1961).

Downing, Paul B., ed., *Air Pollution and Social Sciences* (New York, 1971).

"Drive to Adopt national Water Policy", *Times of India,* 22 July, 1983.

Dubos, Rene, "Man and his Environment", *Britannica Perspectives,* Vol. 1, 1968.

Einstein, A., *My Views,* ed., by S.K. Bandopadhyaya (Calcutta, 1976).

"Environment Research Programme", prepared by NCEPC, Department of Science and Technology, New Delhi.

Forbes, R.J., "The Conquest of Nature and its Consequences", *Britannica Perspectives,* vol. 1, 1968.

Fowler, John M., *Energy and Environment* (New York, 1975).

Fuller, Buckminister, R., *Operating Manual for Spaceship Earth* (New York, 1969).

Gandhi, Indira, "Poverty Greatest Pollution, says Mrs. Gandhi", *Times of India,* 8 September, 1981.

Glenn, Seaborg, "Science, Technology and Development: A New World Outlook", USIS, New Delhi.

Hacoley, Amos H., *Human Ecology* (New York, 1950).

"India Must Develop Own Ecology", *Times of India,* 8 October 1981.

Marion, Jerry B., *Energy in Perspective* (London, 1974).

Misra, K.C., *Manual of Plant Ecology,* New Delhi, 1980.

Mukherji, P.K., *Life of Tagore,* trans, by S.K. Ghosh, 1975.

Mumford, Lewis, "The Future of Cities", in *Basic Issues in Environment,* E.J. Winn, ed., 1972.

Palmslierna, H., *Future Imperatives for Human Environment,* 1972.

Pavithran, A.K., "World Futurology", *Eastern Journal of International Law* (Madras), Vol. 9.

"Plans to Usher India into 21st Century", *Times of India,* 24 October, 1985.

Polunin, Nicholas, "The Biosphere Today", *The Environmental Future,* Proceedings of Ist International Conference on Environmental Future in Finland, ed. by N. Polunin, 1972.

Radhakrishnan, S., *Recovery of Faith,* 1967.

Report on the State of Environment, Prepared by Centre for Science and Environment, New Delhi, 1985.

Sarkar, Mahendra Nath, *The Cultural Heritage of India,* Vol. 1.

Sen, Sudhir, "Blueprint for a Better World", *Times of India,* 2 March, 1980.

The Limits to Growth, A Report to Club of Rome (New York, 1972).

The Mind of J. Krishnamurti, ed. by L.S.R. Vas (Bombay, 1971).

Toynbee, Arnold, "Man and his Soul", *Hindustan Times,* 4 January, 1968.

United Nations List of National Parks and Protected Areas, 1985.

Vivekananda, Swami, *Complete Works,* Vol. II (Calcutta, 1968).

Ward, Barbara and Dubos Rene, *Only One Earth: The Care and Maintenance of a Small Planet,* Report to UN Conference on Human Environment, Stockholm, 1972.

"Wildlife Laws in India", *Times of India,* 4 March, 1985.

Ward, Barbara, *Progress for a Small Planet,* 1979.

INDEX

A

Africa, 2, 22, 123
African Charter on human rights and people rights, 27
Agenda for change, 55-58
    combating deforestation, 56
    conservation of biological diversity, 56
    protecting and managing fresh water, 57
    – – – oceans, 57
    – atmosphere, 55
    sustainable agriculture and rural development, 56
Agricultural organisation, 13
– problem, 136
– regions, 47
– yields, 34
Agriculture biological impact, 41
AIDS, 20
Air pollution, 71
American convention on human rights, 27
Andhra Pradesh, 153, 156
Annual rate of world population growth, 21
Argentina, 77
Asia population, 10
Asian development bank, 101
Atmosphere, 47, 49

B

Bangladesh Rural Advancement Committee, 7
Bangladesh, 7, 99
Bapatla, 154
Beyond recognition, 37
Bhutan, 99
Big firm, 83
Biodiversity, 37-39
Brazil, 77
Bulldozed, 33
Business management, 95
Butane, 86

C

Calcutta, 100
Canada, 76
Carbon dioxide, 49
Caribbean, 11
Catholic world, 8
Catskill mountains, 32
Children, 123
China, 1, 4, 11, 20, 22, 26, 50, 77, 109, 113, 137
Chirala Taluk, 154
Climate change, 49-51
Combining regulations with incentives, 66

Communication facilities, 158
Communities, 69
Consumer of sorts, 76
Consuming the future, 76-78
Consumption bomb, 122-125
    poverty alleviation, 128-133
Consumption, 54
Convention on the elimination of all forms of discrimination against women, 27
– – – rights of the child, 27
Conventional wisdom, 5
Cost-effective, 55
Crucial encounter, 105-107
Cruise ships, 110
Cuba, 26
Cuddapah, 155, 156
Cultivation of nitrogen-fixing crops, 38
Cyclone, 157, 158

D

Demand for food, 3
Democracy, 144
Demographic transition, 2
Denmark, 25
Developing nations, 42
Development, 89-92
    appropriate development, 89
    NGO resolution at the Rio Conference, 91
    stepping backwards, 91
    world way of development, 90
Diarrhoea, 68
Diffuse effects of expanded human activities, 38
Dilemma, 62
Divi, 154
Driving migration, 147-152
    costs and benefits, 149
    question of gender, 149
    refugees, 150
    urban transformation, 148-149

E

Earth temperature, 19
East Godavari, 154-156
Ecological tax reform, 98
Economic to advantage, 102
Economy of A.P., 157
Ecosystems perform service, 30
– our unknown protectors, 30-32
    costly lesson for New York City, 32
    uncertain reaction to climate change, 31
Ecotourism needs, 59
Ecotourism or ecocide, 59-63
    challenge, 63
    industry profits, 61
    local benefits, 60-61
    paying for conservation, 62
    pressure of numbers, 62
Education, 3, 61, 83, 146, 151
Educational efforts, 144
Effects of population growth, 19, 50
Emit oxygen, 30
Employment and poverty alleviation, 134-139
    development gap, 136
    governments role, 139
    minimum wage, 135
    protectionism, 138
    technology and jobs, 137
    – transfer, 135-136

Employment and promoting ecology, 93-98
impacts on employment, 95
increasing resource productivity, 94
production to services, 97
regionalisation of industry, 96-97
waste of resources, 93
Energy and sustainability, 85-88
aftermath of the oil boom, 86
choice of change, 87
– efficiency, 87
Environment burdens, 64
– physical and chemical component, 30
Environmental capital, 17
– goods and services, 16
– protection, 94
– vantage point, 22
Ethiopia, 20, 22, 23
Europe, 34, 53
European Convention for the Protection of Human Rights and Fundamental Freedom, 27
Evidence, 20, 45
Exotic species, 38
Eyes of the public, 102

F

Families, 147
Fast-growing trees, 31
Fertility rates, 1-4
doubling time, 3
fertility rate, 3
population size, 3
Fertilizer production, 38
Fertilizers, 86
Fields of work, 83
Fish, 37
Fisheries, 16, 19
Fishers, 40
Flash floods, 111
Flood water to sink, 153
Food, 40
– production, 140-143
basic facts, 140
investing in people, 141
– supplies, 6, 9
– marketing costs, 143
Forests, 33-36, 52-54
forest area per capita, 34
– cover regulates climate, 34
– products certification, 345
– products, 33
healthy forests boost food production, 33
intergovernmental responses, 35
pressures on forests, 34
technological improvements, 35
France, 109
Frank exchange, 148
Fundamental transition, 41
Future of work, 79-84
dematerialisation, 79
disparities in productivity, 80-81
financial markets that never sleep, 80
haandling information and knowledge, 82
health sector, 82
information technology, 82
leisure economy, 82
Future physical or mental capacities, 106
– population growth, 23

G

GDP, 82, 113
Genetic tests, 105
Genetically correct, 106
Germany, 108
Global demographics, 147
– efforts to slow population growth, 8
– food production, 124
– Infrastructure Fund (GIF), 101
– population, 5, 53
– warming, 125, 44-48
adverse health effects, 45
agriculture at risk, 47
climate changing, 49
growing scientific consensus, 45
rising sea levels, 44
vanishing carbon sinks, 46
Gold-card, 112
Golf courses, 110
Good governance, 69
– news about population growth, 5-9
Government policies, 5
Grainland, 23
Grameen Band, 7
Greek polis, 67
Green peace, 111
– revolution, 40
Gross National Product (GNP), 5, 26, 89, 90, 143
Growth of non-traditional jobs, 83
Gulf, 48, 86
Guntur, 155, 156

H

Habitat, 38
Health, 141, 151
– care, 3
Healthy cities, 68-721
Higher fees, 62
Himalayan dams, 100
Homo sapiens, 32
Hotel and resort development, 114
Households, 147, 69, 151
Housing, 3
– for low income family, 6
– loan, 70
Human activities, 37
– beings, 43, 105
– population, 37
– reproduction, 106
– resources, 81
Hunger, 89
Hydro-power projects, 99

I

Illiterate, 144
Inadequate leadership, 73
India, 1, 4, 20, 23, 69, 99, 127
Indonesia, 33
Industrial nations, 64
– revolution, 85
Institutional framework for credit, 130
– spectrum, 8
Intangible, 79
Intergovernmental Panel on Climate Change (IPCC), 49
– – – Forests (IPF), 35
International human rights treaties, 27
– migration, 148
Invaluable service for the human species, 30
Invisible women, 131
Ireland, 20
Islamic world, 8

J

Japan, 22, 25, 34, 53, 108
Japanese National Doctors Health Insurance Association, 109
Job losses, 137
Jobs, 3, 97
Jungles, 42

K

Karnataka, 92
Kosi High dam, 101
Krishna, 154, 155
Kurnool, 155

L

Lack of money, 105
Latin America, 11, 52
Least developed, 11
Lesson, 1
Level paying field, 132
Living with diversity, 40-43
Local involvement, 60
– populations, 60
Loggers, 40
Long-life good, 95
Lowest status jobs, 150

M

Machilipatnam, 153, 154
Macro-economic tourism, 114
Mahaboobnagar, 155
Major cyclones in Andhra Pradesh, 153-158
Malaysia, 26
Malthus, Thomas, 18, 19
Market segment, 59
Materal mortality, 26
– intensive, 98
Measuring population impact, 16-18
    carry capacity, 18
    ecological footprints of nations, 18
    environmental resource accounting, 16
Medium-level projections, 21
Mexico, 77
Micro-enterprise, 7
Mineral resources, 16
Ministry of finance, 103
Mobilizing, 131
Most likely, 21
Multinational Corporations (MNCs), 92

N

National and international agricultural research systems, 141
– data on life expectancy, 127
– Forest Programmes, 36
– policy, 147
– solutions, 100
Natural resources, 6, 16, 140
Nature of famine, 20
Negative effects, 6
Nellore, 155
Nepal, 99, 100
Netherlands, 25
Nets, 40
Nigeria, 2, 23
NOAA, 47
Non-Governmental Organisation (NGOs), 7, 61, 69, 119, 142
Non-linear organisation, 81
Non-typical jobs, 81
North América, 34, 53
North Atlantic fishing, 125

Northern Europe, 47
Norway, 25
Nutrition, 141

O

OECD, 77, 82
Opportunities, 150
Opportunity costs, 7

P

Pakistan, 2, 23
Part-time work, 83, 98
Peripheral, 127
Persistence of Indian poverty and its alleviation, 126-133
Petrol consumption, 88
Philippines, 137
Pillars, 77
Plastic, 86
Policies, 151
Policy, 67
- makers, 99, 133
Polluters, 109
Poor, 129
- as producers, 130
- people, 75
Population and the environment, 13-15
environment getting worse, 13-14
biodiversity, 14
coastlines and oceans, 14
food supply, 13
forests, 14
fresh water, 14
global climate change, 14
public health, 13
livable future, 14
stabilizing population, 15
taking action, 15
Population challenge, 19-24
- growth, 2, 5, 17, 41, 53
facts and figures, 10-12
- trends, 1
Poverty, 129
Prakasam, 155
Problem(s), 127
of co-ordination, 157
- employment, 134
Progress, 127
Promoting energy saving programmes, 86
Propane, 86
Pro-poor tourism, 113-121
categories of people, 116
cultural impacts of tourism can be positive or negative, 117
effects on the livelihoods of the poor, 116
policies to enhance
enhance economic opportunities and a wide range of impacts, 118
incorporate pro-poor tourism approaches into mainstream tourism, 120
multi-level approach, 119
pro-poor tourism, 118-121
put poverty issues on the tourism agenda, 118
reform decision-making systems, 120
work through partnerships, including business and tourists, 119

positive development impacts of tourism, 117
potential of pro-poor tourism, 115
tourism and aid, 113
Prosperity, 78
Public commitments, 42

Q

Question, 18

R

Rank-and-file employees, 81
Rao, N.T. Rama, 157
Rapid growth, 62
– population growth, 10
– urban growth, 147
Rapidly growing population affect the environment, 17
Rayalaseema, 155
Reducing maternal mortality, 25-26
Refugee-life- situation, 150
Religion, 144
Repalle, 154
Replacement, 4
Replicable, 31
Rightful share, 78
Rio Conference on Environment and Development, 74
Rising global temperatures, 45
River Indo-Nepal, 100
River Karnali, 101
Role players, 63
Royal Caribbean cruise, 110
Rural sector, 148
Russia, 47
Rwanada, 20

S

Safe motherhood is a human rights issue, 27-29
application of human rights, 29
implementation of laws, 29
reform of laws, 28
rights relating go
equality and non-discrimination, 28
foundation of families and of family life, 27
healthcare and the benefits of scientific progress, 28
– relating to life, liberty and security of the person, 27
Safety net, 127, 130
Schumachar, 92
Science, 8
Service life, 96
Sexually active young people, 7
Sikkim, 60
Skilled workers, 97
Sluggish markets, 93
Small is beautiful, 92, 136
Smaller families, 20, 24
Social activities, 99
– economy, 70
– problem areas, 95
Socio-cultural intrusion, 117
Socio-economic problem, 134
Sough Asia, 6
Source of mechanical energy, 85
South America, 113
South Asia, 2, 11, 145
quarrels over water, 99-102
South East Asia, 113

Southern Africa, 11
Spanish Island Minorca, 111
Standard oil, 85
State forests, 124
Storing freshwater, 31
Stricken area, 157
Structural change in economies and societies, 131
Sub-Saharan Africa, 6
Sudan, 20
Supply food, 158
Sustainable cities, 72-75
  disaster mitigation relief and reconstruction, 74
  incorporating environmental concerns, 74
  increasing awareness of gender issues, 74
  reducing poverty and creating jobs, 73
Sustainable tourism and the environment, 108-112
  need for action and education, 111
Synthetic rubber, 86

T

Tatopani, 60
Technologies, 105
Technology, 8, 18, 90
Tourists, 109
Traditional agriculture badly needs, 41
– agroecosystems, 41
Traffic problems, 66
Trickle-down, 126, 128

U

UN food and agriculture organisation, 52, 13
UN medium projection, 122
UN population, 20
  division, 122
  projection, 20, 24
Underdevelopment, 89
United States, 23, 32, 77
Urban areas, 42
Urban environmental problem, 66
– population growth, 72, 73
Urbanisation and the environment, 64-67
  changed urbanisation pattern in India, 64
  growing environmental damage, 65
  innovative approaches to solutions, 65-67
US Department of Energy (DOE), 49
US State Department, 110
USA, 25
Use economics, 102-104
  international organisation, 103
  promising initiative, 103

V

Venezuela, 77
Visakhapatna, 154
Vital function, 31

W

Warmer global temperatures, 45
Waste, 89
Water pollution, 71
Water tables area, 19
Water, 40
Weakness, 134
Weedkillers, 86
West Godavari, 154
Widespread food insecurity, 140

Wildlife, 16
Willingness-to-pay surveys of ecotourists across, 62
Wishes, 129
Women, 123, 150
– are educated, 145
– status, 145, 150
– work, 132
Wood, 40
World Bank, 34, 101
created a Tourism Projects Department, 113, 114
– forest cover is shrinking, 33
– growing population, 55
– Meteorological Organisation (WMO), 45
– – , 11, 19, 34, 72, 122, 141
growth, 4, 23
– Tourism Organisation (WTO), 108
World War II, 86
World War III, 5
World-wide rate of population growth, 10
WWF, 111